Denis P. da Silva
Josi M. de B. Santos
Maria de N. M. Maciel

Morphometric Characterization of the Microbasins of the Apeú River, Pará, Brazil

Denis P. da Silva
Josi M. de B. Santos
Maria de N. M. Maciel

Morphometric Characterization of the Microbasins of the Apeú River, Pará, Brazil

Using Digital Elevation Models and Multivariate Analysis

ScienciaScripts

Imprint

Cover image: www.ingimage.com

This book is a translation from the original published under ISBN 978-613-9-67959-1.

Publisher:
Sciencia Scripts
is a trademark of
Dodo Books Indian Ocean Ltd. and OmniScriptum S.R.L publishing group

120 High Road, East Finchley, London, N2 9ED, United Kingdom
Str. Armeneasca 28/1, office 1, Chisinau MD-2012, Republic of Moldova, Europe
Printed at: see last page
ISBN: 978-620-8-18936-5

SUMMARY

INTRODUCTION

The first characterization of water basins as an environmental system was described in 1962 by Richard John Chorley, who is considered a pioneer in the use of this concept. Since then, the evolution of technology and the development of new scientific methods have focused on understanding this system, seeking precise results to improve the exploitation of the environment. The more they are studied, the more other angles of vision can be observed in this process, generating new questions related to various areas of knowledge.

When it is necessary to use models to manage water resources, environmental systems are the basis of reference, due to their interactions with the environment, where inputs, outputs and transformations of the water cycle occur. The use of this system is based on the legislation of the National Water Resources Policy (PNRH), in which river basins are identified as the territorial unit. The need for a more conscious use of water has led to an increase in research in this area, generating better instruments for public authorities in order to introduce sustainable development.

The study of the Earth, supported by technology, makes geoscience one of the fields involved in the characterization of river basins, as well as the interaction with computer software and statistical analysis. The combination of these areas of knowledge and technology led to the consolidation of remote sensing in the 1980s and the advent of the Geographic Information System (GIS), with computer interfaces integrated with the data obtained from the relief, with the aim of modeling the processes that occur according to their specific scales.

Morphometry is a branch of geomorphology that analyzes the components of river basins, using satellite images and maps for physical characterization. Therefore, the geoprocessing of data through GIS makes the diagnosis of the area a faster and more complex process, providing a model that is more similar to reality and at a lower cost.

Thus, topographic elevation data of the study area is required in order to generate the digital terrain model (DEM) that best represents the site. From this, the study parameters that need to be interpreted are extracted. The generation methodology, the way in which the data is evaluated and the statistical treatment applied define the quality and degree of approximation (or distance) of the results in relation to reality.

Statistics is one of the pillars of watershed modeling analyses. Multivariate analysis is used to group together the **n** variables generated by the watersheds, allowing them to be analyzed simultaneously. The use of this statistic synthesizes, orders and separates the variables in relation to their similarities (or dissimilarities), making it possible to optimize the information contained in the input data. In this way, structural patterns in natural environments can be detected by analyzing the geomorphological behavior of the hydrographic basin and identifying homogeneous zones for land management and planning.

OBJECTIVE

General objective

The aim of this work is to apply multivariate analysis to morphometric parameters extracted from the automatic delimitation of micro-drainage basins, with a view to characterizing them as management units in order to provide subsidies for the territorial management of the basin.

Specific objective

- Model the drainage network, as well as the hierarchical micro-basins that make up the Apeù River basin using a hydrologically corrected Digital Elevation Model (DEM);
- To carry out the morphometric characterization of the drainage basins using the multivariate statistical technique of Principal Component Analysis;
- Identify, by means of cluster analysis, watersheds with similar morphometric characteristics.

LITERATURE REVIEW

Geomorphological analysis

According to Florenzano (2008), geomorphology is the science that studies landforms according to their genesis, composition and the processes that act on them. It uses remote sensing and GIS technologies to improve this science. Geomorphology is subdivided into objects of study, with morphology being one of the most researched due to its relationship with the environment and land use. Within this field of study, morphometry is used to characterize the relief in a quantitative way, through morphometric indices.

Fonseca (2010) points to the Geographic Information System as a highly relevant tool when it comes to characterizing elements and values, as well as synthesizing data and composing profiles of the earth's surface. Therefore, geomorphology benefits from the use of this technology, which has been improved over time, and can thus support decision-making and the resolution of questions involving environmental issues of land use.

Morphometry analyzes variables related to relief in a quantitative way. Among them, altitude, altimetric amplitude, slope extension, declivity, drainage density, river frequency and interfluvial amplitude are pointed out by Florenzano (2008) as being the most used for geomorphology research, all of which are related to the environment and its fragility in 3

the human being. Each of these parameters will be specified in the course of this text, always returning to the central objective. However, it is worth noting that for Fonseca (2010), it is necessary to add the prefix GEO in front of the word morphometry to emphasize the shape that is modeled by geomorphology from that found in nature.

The river basin as a territorial management unit

The delineation of a watershed requires the definition of a watercourse, a reference point or section along this watercourse, its characteristics can be analyzed in relation to the relief, type of soil, vegetation present, use and management of the area, Tucci

(1993). The main river of a basin is given by the longest line of water flow linking the source to the outlet, Bernardi et. al., (2013). The outflow is the place where the precipitated water in the basin converges, i.e. it is located in the lowest part of the relief, Braga et al., (2005). Currently, the most widely used way of analyzing river basins is through GIS.

Due to the characteristics that make up a watershed, it can be understood as a system, where it is necessary to obtain a holistic and punctual understanding of its dynamics, so that it is possible to establish a diagnosis and define its situation at that time. From this knowledge, it is possible to describe the basin's aptitudes and to encourage and/or suppress action in it. This demonstrates the direct influence of watershed management in relation to the management of the territorial basin as a planning unit.

Collares (2000) emphasizes in his text the importance of using river basins as boundaries for planning units. The countries that pioneered this form of boundary dimensioning are France, Greece, the United Kingdom and the United States. It was introduced in Brazil in 1976 and began in 1987 in the state of Sao Paulo. The southern and southeastern states of Brazil were the pioneers when it came to planning the implementation of the PNRH.

In Brazilian legislation, river basins are defined as a territory for public policies and a unit for managing water resources, i.e. a basic management unit, Bernardi et al., (2013). On January 8, 1997, the Federal Constitution enacted Law No. 9.433, referring to the National Water Resources Policy (PNRH), in order to establish parameters that characterize a body of water and how it should be cared for and used, seeking to ensure future generations such heritage. The decree establishes the instruments of the PNRH, and Section I, Article 8 establishes the preparation of the PNRH for river basins in relation to the state in which it is located.

The morphometric characteristics of a body of water are what classify a hydrographic basin, but it can be subdivided into sub-basins. Carmo and Silva (2010) point out that these subdivisions help make it easier to locate diffuse problems in micro-basins, such as areas of natural resource degradation that could compromise the sustainability of the basin. Each basin and/or micro-basin plays an extremely important role in the occupation of the territory, which defines different drainage patterns and can result in a unit that is easily managed by local societies or actions that compromise the use of the site, Rocha and Vianna (2008).

Rocha and Vianna (2008) emphasize in their text the damage that environmental degradation causes, the imbalance of the natural landscape forms an accumulation of environmental problems sectored in the integrating agent of the natural and social, the hydrographic basin. With the increase in population and the expansion of water exploitation, this natural resource has been deteriorating. However, through the implementation of the PNRH, the aim has been to ensure that man's actions on the banks of the basins are orderly and that damage is minimized. Therefore, for future uses of these basins, prior studies are needed to show the socio-environmental panorama of the locality, in accordance with the interests, guaranteeing the quality and quantity of the natural resource.

Morphometry of micro-watersheds

More objective indices and parameters for the quantitative study of drainage systems were introduced by the hydrological engineer Robert E. Horton in the mid-1940s, who initiated a more precise watercourse planning system (HORTON, 1945). From the interaction of the drainage network and the relief characteristics of a given surface, morphometric models can be generated, from which quantitative data can be extracted that can help in the understanding of river dynamics, sediment and energy transport, and these models allude to a basic unit of planning and analysis: the hydrographic basin.

On the European continent, the first attempt was made to classify watercourses into a system of drainage basins. In this classification, the widest, most branched and the main watercourse were always ordered as 1 (one), and the smallest streams were ordered in ascending order. However, Horton (1945), feeling that the main river should be the one with the highest order and the smaller unbranched tributaries should always be designated by the same number, used a drainage system that was the inverse of the

European system. [a]In this system, those that have no tributaries are always designated as F-order; 2-order canals only receive F-order tributaries; 3-order canals can receive one or more 2-order tributaries, but they can also receive F-order tributaries; 4-order canals receive 3-order and generally lower-order canals, and so on.

However, the author created some rules to determine which is the tributary and main channel from the last bifurcation: 1) starting downstream of the confluence, extend the line of the watercourse upstream, beyond the bifurcation, following the same direction. The confluent channel with the largest angle is the lower order; 2) if both courses have the same angle, the river with the smaller length is generally designated the lower order. The process of re-numbering should be carried out at each confluence with a higher order, until the nth order channel extends from the confluence to the source of the longest tributary (CHRISTOFOLETTI, 1980).

Due to some subjective decisions in Horton's system, many researchers began to adopt a different system, which was introduced by Arthur N. Strahler in 1954. Strahler, in 1952, which consists of: channels without tributaries are considered to be first order, extending from the source to the confluence; second order channels arise from the confluence of two first order channels, and only receive first order tributaries; third order channels arise from the confluence of two second order channels, and can receive second and first order tributaries; fourth order channels arise from the confluence of two third order channels and can receive tributaries of lower orders, and so on. The new system proposed by Strahler eliminates the concept that the main river should have the same order number throughout its length and the need to re-number at each confluence (CHRISTOFOLETTI, 1980).

The river hierarchy system initiated by Horton enabled the author to formulate some laws that dictate the ideal behavior of river basins, and these were presented by Christofoletti (1980), which can be summarized as follows:

- **Law of the number of channels:** proposed by Horton (1945), in a given basin, the sum of the number of channels of each order forms an inverse geometric series, the first term of which is the first order unit and the ratio is the bifurcation ratio.

o **Law of channel** lengths: proposed by Horton (1945), it says that in a given basin, the average lengths of channels of a given order are ordered according to a direct geometric series, where the first term is the average length of the first-order channels, and the ratio is the relationship between the average lengths.

o **Law of channel slope**: proposed by Horton (1945), in a given basin there is a defined relationship between the average slope of channels of a certain order and that of channels of an immediately higher order, which can be expressed by an inverse geometric series, in which the first term is the average slope of the first order quays and the ratio is the relationship between channel gradients.

o **Channel basin area law:** proposed by Schumm (1956), it says that in a given hydrographic basin, the average area of the drainage basins of the channels of each order is ordered approximately according to a direct geometric series, in which the first term is the average area of the first order basins.

Cherem (2008) also mentions that these laws were set up by Horton and Schumm in order to characterize a pattern of normality in the behaviour of basins, helping in the methodologies of representing and structuring them in order to verify that they are close to the established "patterns of normality". These laws opened the door for other researchers to develop other mathematical parameters to mathematically represent the evolution and configuration of river basins: Strahler (1952), Schumm (1956), Strahler (1957, 1958), Chorley (1962), and Chorley and Kennedy (1972), are some of them.

The attributes of the watershed used to compose the morphometric parameters are: area; perimeter; and length of the watershed; length of the river channels; vector length (valley length) of the river channels; number of river channels; slope of the watershed (minimum, average and maximum); slope of the channels; and altimetry (minimum, average and maximum). The parameters correspond to the correlations between these attributes and the relief of a basin, and are grouped into linear, zonal and hypsometric. Linear parameters deal only with the attributes of the drainage network; zonal parameters deal with the attributes of the drainage network and the non-hydrographic areas of the relief (slopes and tops); hypsometric parameters deal with the relief, as well as correlating it to the drainage network and the hydrographic basin. It is important to emphasize that in order to represent the hydrographic basin for morphometric analysis, it needs to be generated using a consolidated methodology, because it is based on its measurable elements of analysis - its area, drainage network and the arrangement of slopes (relief) - that the morphometric parameters of the basin will be generated, which will indicate its possible behavior (CHEREM, 2008).

Technological development, combined with geomorphology, allows for a high availability of data and techniques to be applied in remote sensing, enabling the study of maps and analysis of measurements of geomorphological variables 8

quickly and accurately (Florenzano, 2008). The authors Yunus, Oguchi and Hayakawa (2014); Li (2014); Hosseinzadeh, (2011) and Gunal and Guven (2015), use the Digital Terrain or Elevation Model (DEM) as an important tool for capturing topographic and geomorphological parameters to generate informative maps derived from algorithmic analysis in a GIS environment. The images that make up the DEM must undergo georeferential preparation of each variable, the most commonly used example being the Digital Elevation Model (DEM) (VALERIANO (2005) and FLORENZANO (2008)).

EAW images can be obtained from topographic maps, orthophotos and/or images from orbital sensors which, together with the help of GIS, have led to the development of methods for extracting the variables needed for processing and insertion into the computer environment (VALERIANO, 2002). Florenzano (2008) adds to this development the availability of *Shuttle Radar Topographic Mission* (SRTM) data for South America, which has given a boost to studies using multi-dimensional DEM, combined with the lower cost of this automatic method compared to the traditional one.

The data from the SRTM project was collected in 2000 through a contribution from Nasa *and* the *National Imagery and Mapping Agency* (Nima), with the aim of generating a worldwide MDE. The data for South America was released for free access in 2003 with a resolution of 90 meters, but after some analysis, interference was observed which required precise techniques to remove it without altering the material

(FLORENZANO, 2008). The Topodata database uses the data provided by the SRTM and applies standardized processing, with technical specifications and a unified refinement, in order to generate basic information on the geomorphometric variables of a site, ready to be used in GIS's, however this material does not satisfy when geomorphological analyses are required (VALERIANO; ALBUQUERQUE, 2015).

Analyses carried out on micro-watersheds generally focus on problems such as erosion and/or flooding, through direct estimates and mapping of the factors involved, however, consideration should be given to 9

that each type of relief requires a different processing method, so it is recommended that there is a stratification of the processing of these data within the scale worked on, providing a segmentation in topographic environments (VALERIANO, 2002). Examples of work on geomorphological analysis in a GIS environment include: Valeriano and Moraes (2001), Valeriano (2005), Sodré et al., (2007), Santos and Morais (2012), Soares and Souza (2012) and Veiga et al., (2013).

These studies prove the feasibility of using morphometric parameters to analyze river basins. Given the great contribution of these techniques to regional geomorphological analysis, it is essential that morphometric parameters are generated using standardized methodologies. Thus, the parameters used to represent the basin in a virtual environment can achieve the behavioral patterns of the hydrographic basins established by Horton's laws, and the degree of their development can thus be seen.

Multivariate data analysis applied to the morphometric study of watersheds

When morphometric analysis is applied to micro-basins by river hierarchy, the volume of data grows absurdly. For example, if the number of parameters used for the analysis is nine, this will be multiplied by the number of micro-basins for each river order. As a result, it is necessary to investigate the interrelationships between the micro-basins using multivariate statistical procedures due to the large volume of data generated.

Multivariate analysis refers to the set of exploratory, descriptive and inferential statistical techniques used to analyze situations involving a large number of variables simultaneously. The techniques used in multivariate analysis aim to synthesize, order or separate the set of variables in an optimized way, enabling better detection of structural patterns in natural environments (ROCHA, 2015).

Among the various techniques applied in multivariate statistical analysis, Oliveira (2008) highlights two techniques applied in the study of drainage basin morphometry: principal component analysis (PCA) and grouping or *cluster* analysis (AA).

According to Vicini and Souza (2005), the aim of *PCA* is to describe a large set of data with *p* variables related to *n* individuals - in the case of this study, the variables will be the morphometric parameters and the individuals will be the various watersheds. By applying the technique, the number of variables is reduced, not by simply selecting certain variables, but by constructing new synthetic variables obtained by linearly combining the initial variables. The method also makes it possible to identify the measures responsible for the greatest variation between results, without significant loss of information. The new variables generated are called principal components (PC), which are equivalent to the measures already mentioned. It is worth noting that the reduction is only possible if there are nonzero correlation coefficients between the initial variables.

Cluster analysis studies a whole set of interdependent relationships, making no distinction between dependent and independent variables. The purpose of this technique is to gather individuals observed in the data set into groups with homogeneity within the group and heterogeneity between the groups, in order to propose classifications between the individuals based on the set of input variables. When viewed in this way, cluster analysis is the inverse of principal component analysis in that it reduces the number of individuals, rather than the number of variables, concentrating them in a much smaller number of groups. In this study, CA will be applied to the PCs of the micro-basins, with the aim of identifying by grouping groups of basins with similar characteristics so that management groups can be formed in the basin to better organize its territory, taking into account the morphometric characteristics of the Vicini and Souza (2005) micro-basins.

CHARACTERIZATION OF THE STUDY AREA

The area selected for the work was the hydrographic basin of the Apeù River (Figure 1), located in the state of Parà, between 3 (three) municipalities in the northeastern mesoregion, which is concentrated 70% in the municipality of Castanhal, 20% in Inhangapi and 10% in Santa Izabel do Parà. The Apeù river rises on the Buriti farm, in the municipality of Castanhal, and flows into the Inhangapi river (SANTOS, 2006).

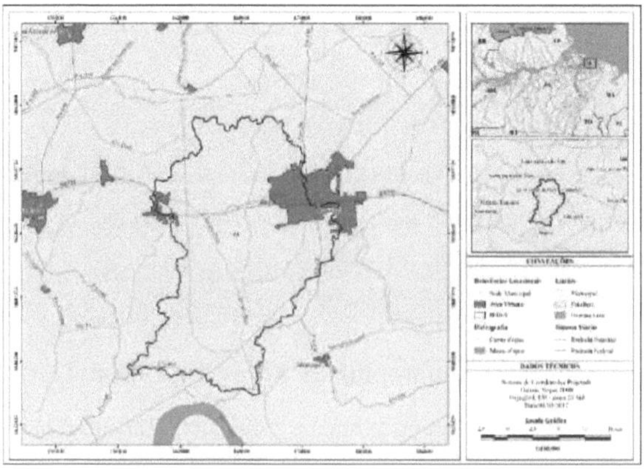

Figure 1 - Location map of the Apeù River Hydrographic Basin (BHRA), located in the northeastern mesoregion of Pará.

Source: author

Geologically, this area is made up of the Tertiary soils of the Barreiras Formation, as well as recent Quaternary sediments, represented by gravels, sands and unconsolidated clays, which occur in narrow, discontinuous strips following the watercourses. The geomorphological features are characterized by the presence of hills with flattened and moderately dissected tops, making up one of the sectors of the Lower Amazonian Plateau, and the Pleistocene and Holocene sedimentary plateau. The types of soil that occur in the basin (figure 2) are predominantly two: typical and argissolic Yellow Dystrophic Latosol and typical Red-Yellow Dystrophic Latosol (SOUZA et al., 2012).

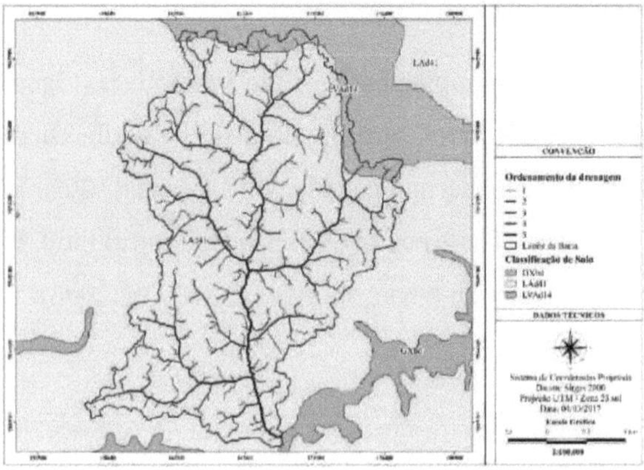

Figure 2 - Distribution map of the soils present in the BHRA. Where: GXbd (Gleissolo Hàplico Tb Distròfico tipico); Lad41 (Latossolo Amarelo Distròfico tipico e argissólico) and LVAd14 (Latossolo Vermelho-Amarelo Distròfico tipico). Color scheme according to IBGE (2008).

Source: author.

The vegetation cover is dominated by Dense Ombrophilous Forest or Dense Terra-Firme Forest, which can be seen bordering the rivers, where it becomes characterized as Igapò Forest or Seasonal Igapò Forest. To a lesser extent, vegetative formations popularly known as capoeiras are also found, originating from human intervention throughout the occupation of the region (SOUZA et al., 2012).

MATERIAL AND METHODS

The work can be divided into the following stages: 1) digitization of the drainage by means of visual identification based on photointerpretation techniques; 2) construction of the hydrologically corrected digital elevation model (DEM); 3) delimitation of the drainage micro-basins and automatic extraction of the hydrography; 4) calculation of the morphometric parameters of the micro-basins; 5) multivariate statistical analysis of the morphometric parameters of the drainage micro-basins.

The *software* used to generate all the data in this study was: Arcgis 10.3 and Qgis 2.14 for the analysis and processing of spatial data; BioEstat 5.0 for data transformation and principal component analysis; Microsoft Excel for tabulating the data and generating graphs; Action Stat 3.1, which is an add-in for MS Excel, was used to carry out the cluster analysis.

Digitization of the drainage network

To identify the area covered by the drainage vectorization, the following were used: topographic maps of Belém and Castanhal - SA-22-X-D-III of Belém (**index map** - mi384) and SA-23-V-C-I of Castanhal (**index map** - mi385), both on a scale of 1:100.000 scale, acquired from the spatial database of the Army's Geographic Service Directorate (DSG), as well as sheets 01S48_ZN and 01S495_ZN from the TOPODATA MDE (project carried out by the National Institute for Space Research - INPE). Once the main course of the Apeù River had been identified on the topographic maps and the approximate area of the hydrographic basin visually identified on the MDE, as can be seen in Figure 3, the polygon of coverage was defined for the vectorization of the drainage. This vectorization was done using photointerpretation techniques from RapidEye satellite images.

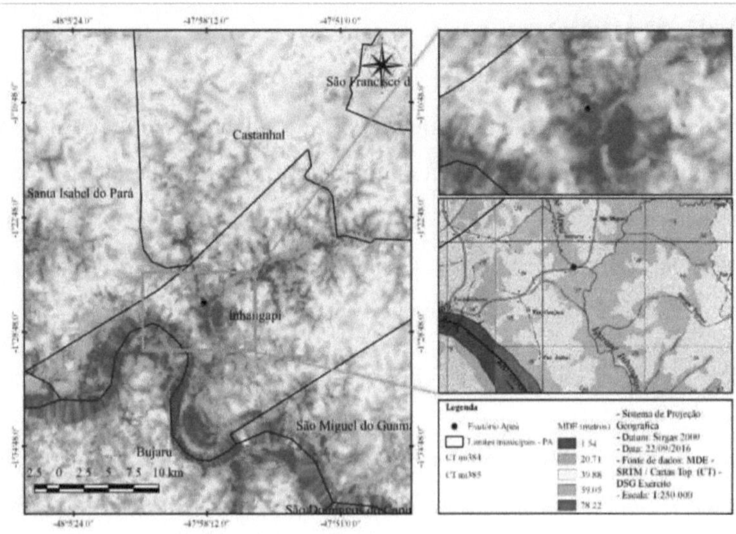

Figure 3 - Method of identifying the Apeù River Basin.

Source: author

The RapidEye satellite images were acquired from the geocatalago (digital collection of satellite images from the RapidEye constellation) on the website of the Ministry of the Environment (MMA) via the institutional *login of* the Federal Rural University of Amazonia. The study area is covered by four satellite scenes, whose grid identifications are: 2238428, 2238528, 2338401, 233850. These images underwent pre-processing in GIS *software*, which included: mosaicking the four scenes that cover the basin area; cutting out the mosaic (for the polygon that covers the drainage vectorization) and a false color composite R5G3B2, which was chosen because it allows for better identification of the drainage and riparian forest. It is worth noting that the images acquired are framed at level 3A[1] of processing, thus eliminating the need for the user to register the image. Based on this treatment, the drainage was vectorized in the image referring to the study area, and the image has a spatial resolution of 5 meters, allowing for a mapping scale of 1:25,000. In addition to the Ibtointerpretation techniques used for vectorization, the drainage from the topographic maps was also used to guide the

1 On this site you can find the characteristics of the Rapideye image at level 3A: https://www.planet.com/products/#satellite-imagery

identification of the hydrography.

After vectorization, the vector drainage network was refined using images from Google Satellite[2] , as the RapidEye satellite's limited spatial resolution of 5 meters did not allow for the identification of less significant drainages.

Preparation of the hydrologically corrected MDE

Hydrological algorithms generally produce significant results on steep slopes, but there are difficulties when there are relatively flat areas. Thus, the greatest effort has been made to delineate water flow in relatively flat areas, where the river channel is very small and the absence of a gradient prevents computational delineation. Because of this, there are many errors in the automatic delineation of channels and consequently drainage basins from untreated MDEs, since information on the location of the drainage network is poor. In order to meet this need, this work uses a methodology adapted from that proposed by Oliveira (2008). The methodology consists of altering the elevation values of the pixel through which the vectorized drainage network passes. This made it possible to introduce water flow and obtain a more accurate model of the hydrographic structure and, consequently, a better delineation of the drainage micro-basins.

MDE pre-processing

Before applying any procedure to the Topodata MDE, it was necessary to pre-process the image in a GIS environment. Therefore, as soon as the two scenes were acquired, the mosaic was made and then reprojected to fix the cartographic reference system, being reprojected to the Sirgas 2000 datum.

When working with micro-watershed delimitation, it is recommended to use topographic data (MDE) in flat coordinates. In this work, the coordinate system used was the Universal Transverse Mercator (UTM). The UTM coordinate system is characterized by the fact that the globe is divided into 60 spindles, and in some situations the cartographic boundary being worked on is located between two spindles,

---

2 The images used were those available in the *OpenLayers plugin*, an add-on to the Qgis *software*.

which is what happened with the BHRA, so it was decided to assign it to the spindle in which most of the basin is located, i.e. spindle 23. Therefore, the mosaic in the Sirgas 2000 geographic system was reprojected into the UTM spindle 23, Sirgas 2000 system.

Once the mosaic had been reprojected, the *raster* extraction process was carried out by means of a vector mask, or clipping, this mask being an area defined by the author to reduce the volume of data for processing, given that the location of the basin had already been identified, as mentioned at the beginning of the methodology. It is important to note that all the data used during the process was in the same projection (Sirgas 2000 / UTM Zone 23 South).

Once the area of interest had been processed, the data was rescaled from 32 bits to 16 bits, making processing less costly.

In summary, the procedures for preparing an EAW for micro-basin delimitation were: 1) download the Topodata EAW; 2) mosaic all the Topodata sheets 3) fix the projection of the mosaic onto the geographic system; 4) reproject the mosaic onto the flat system; 5) reproject the polygon for clipping onto the flat system; 6) cut out the mosaic using the polygon as a mask and 7) transform *Floating Point* 32 *bits* to *Unsigned Integer* 16 *bits*.

Hydrological correction of the EAW

The procedure to ensure the demarcation of the drainage network consists of deepening the hydrography in the EAW in order to introduce the flow direction along the channels. This methodology can be divided into the following stages: 1) vectorization of the drainage network using photointerpretation of satellite images; 2) extraction of the topographic elevations from the EAW in point vector format; 3) vector operations between the elevations and the vector drainage network; 4) interpolation of the new vector data to generate the hydrologically corrected EAW; and 5) calculation of the flow direction.

The vectorized drainage based on the satellite images was used as the vector for selecting the in-depth topographic elevations. Therefore, a vector grid was extracted from the pre-processed MDE raster in polygon format and in the size of the resolution of the input data (30 x 30 meters) and the centroid of each polygon was then extracted in vector point format. The point mesh was used to collect the elevation information, in meters, from the MDE by means of a GIS software processing module. Consequently, each point represents a regularly distributed topographic elevation at an interval of every 30 meters.

Once the drainage area had been vectorized, a spatial selection query was carried out so that the vector intersected the regular mesh of the topographic elevation of the area of interest, within a 30-meter interval, on both sides of the drainage line. For the selected elevations, a "yes" code was assigned in the attribute table field, and for those not selected, a "no" code. Using this code, only the pre-established elevations were selected, which were subtracted 15 meters from their elevation values and then a constant value of 50 meters was added to the result of the previous operation for all elevations - the constant is intended not to leave negative elevation values, given that the minimum elevation identified in the MDE was 2 meters. The aim of this operation was to ensure that the processing of the drainage extraction and the respective micro-basin preserves the characteristics of the vectorized drainage as much as possible, so that the generated drainage micro-basins are well modeled.

Once the vector data had been properly processed, it was used as *input* to the *Topo to Raster* interpolation algorithm in ArcGis 10.3 *software*, which differentiates the data according to its type - it represents: elevations as *PointElevation*; drainage as *stream* and the *boundary as Boundary* - to generate the EAW.

It is important to note that the hydrologically corrected model was used exclusively to generate the drainages and drainage micro-basins according to their order. For the altimetry and slope calculations and maps, the pre-processed Topodata model was used, with only the depressions corrected using the TauDEM extension in ArcGis 10.3. The scale of analysis allowed by the Topodata MDE is 1:100,000, as mentioned by Valeriano and Rossetti (2010). With the refinement, the model's drainages and micro-basins became closer to reality, but the information content and positional quality did not increase, as this is a characteristic inherent to the moment of imaging of the SRTM (*Shuttle Radar Topography Mission*) sensor.

Also according to Valeriano and Rossetti (2010), the refinement of the SRTM, which is the Topodata product, can favor more detailed analysis of the relief, and this already satisfactorily attends the study, since, according to Florenzano (2008), morphometry - the object of study of the research - refers to the quantitative aspects of the relief.

Delimitation of micro-drainage basins

The methodology used to represent the watershed in a Geographic Information System (GIS) environment was based on the works of Tarboron (1997) and Tarboton and Mohammed (2013).

The boundary of the hydrographic basin obeys topographic parameters, where an imaginary line is drawn through the highest points of the terrain, crossing the watercourse only at the outflow, according to Christofoletti (1980). The outflow is demarcated at the confluence with another watercourse.

The method for dividing drainage basins was based on fluvial hierarchization, first proposed by Horton (1945), and later Strahler (1952) defined a simpler and more functional method, which is still used today.

It is worth pointing out that the order of the drainage micro-basins is the same as that of their respective channels, so each segment of a certain order is responsible for draining an area. In the case of second-, third- or higher-order micro-basins, the area subordinated to them also includes the area of all the lower-order segments that are subsidiary to it. Thus, a second-order basin includes the areas of the first-order segments, as well as the specific areas that drain into the second-order segments; a third-order basin includes the areas of the second-order basins that are subsidiary to them, the first-order basins that flow directly into the third-order channels, as well as the specific areas that drain into the third-order channels; a fourth-order basin encompasses the third-order areas that are subsidiary to it, the areas of the second- and first-order basins that flow directly into the fourth-order channel, as well as the specific areas that flow into the fourth-order channel; and so on (CHRISTOFOLETTI, 1980).

The *Tarrain Analysis Using Digital Elevation Models* - TauDEM (TARBOTON; MOHAMMED, 2013) algorithm can be used to computationally obtain the drainage micro-basins according to the fluvial hierarchy proposed by Strahler. The input data for this algorithm is the DEM. From this, the algorithm generates: the MDE with the removal of depressions, the flow direction grid using the D8 method of the drainage network (TARBORON, 1997), slope, and contributing area. With this data, TauDEM can delimit the drainage micro-basins and their respective drainage network.

In this way, the TauDEM method generated line segments according to the fluvial hierarchy. Thus, the automatic delimitation of the drainage micro-basins takes place for each segment of the ordinary channel, generating a polygon for each segment. This made it possible to analyze the behavior of the drainage micro-basins by river order.

After pre-processing and hydrological correction of the MDE, the generation of the drainage basins and hydrographic network began using the TauDEM algorithm.

The MDE as *input* to the *pit* removal module generated a filled raster, i.e. all the cells with a lower value than those around them with a higher elevation are corrected to a higher value considering the neighboring cells, in order to eliminate these "*pits*" in the image.

The result of the previous module is the input data for mapping the direction of flow using the D8 method, which estimates the direction of water flow in eight directions (TARBOTON, 1997), using the *D8 Flow Direction* module. The result is an image with pixels with values ranging from 1 to 8, in integers.

The mapping of the contributing area corresponds to the boundary areas of the basins from the accumulated flow direction (module *D8 Contributing Area*), in this product it is already possible to see the shape of the basins that occur in the area. This is the first version of the contributing area, showing the existing sub-basins.

With the *Stream Definition by Threshold* module, a raster of the drainage network contained in the first version of the contributing area (not the drainage network of interest in this work) was extracted based on a threshold - the lower the threshold, the denser the resulting drainage. A threshold of 100 was used to identify the outflow point (which is the point to which the entire water load of the basin is drained, from downstream to upstream of the basin) of the drainage basin of interest.

A *shapefile* was created to represent the outflow and entered as a parameter in the *D8 Contributing Area* module to generate the cumulative flow direction for the basin of interest only, thus generating the second version of the final contributing area.

With only the study basin as a contributing area, it is possible to extract the drainage for it using *Stream Definition by Threshold*, which will be the second version. In this process, the drainage was defined by a threshold of 300, as this was what generated the network closest to the hydrography vectorized by photointerpretation. The aim of correcting the MDE was to demarcate the real hydrography so that the automatic extraction of the network and basins would preserve these real channels as much as possible.

The *Stream Reach and Watershed* module was responsible for transforming the raster drainage into a vector format with the river hierarchy, as well as delimiting all the individualized micro-basins for each channel segment. The micro-basins come in raster format, so it was necessary to polygonize this data.

When TauDEM generates the hydrography and micro-basins, they are fragmented, so that at each confluence, a new stretch of channel appears along with its drainage basin. For this reason, the fragments were joined by information on the order of the channels and the micro-basins were joined according to their order, which was done manually, following the criterion that the first order basins are subsidiaries of the second order basins; while the first and second order basins are subsidiaries of the third and so on, as mentioned by Christofoletti (1980). During the hierarchization of the micro-drainage basins, fragments were noticed that did not fit into the order now hierarchized.

With the drainage and micro-basins in vector format, it was possible to calculate the variables and then generate the parameters. When TauDEM processes this data, it generates a code that is common to both pieces of data, making it possible to relate the variables calculated in the drainage to the basin data and vice versa. The code represents the identifier of each basin and respective drainage linked to all the values of variables, parameters and order.

Definition of treatments for analysis

As mentioned above, during the hierarchization of the drainage basins, we noticed fragments that did not fit into the order now hierarchized. It was therefore decided to carry out the analysis with and without these fragments. This was done in order to check whether the basins that only have the order analyzed (without fragments) would be sufficient to define a distribution pattern for the parameters, components and groups, or whether the fragments that are not part of the order analyzed would also be important to complement the areas without information.

In view of the above, the treatments described in Table 1 were used to carry out the analyses.

Table 1 - List of treatments for analysis.

index	Treatment	Description
1	BO1	Only for 1ª order watersheds analyzed
2	BFOl	Micro-watersheds of 1ª order plus fragments
3	BO2	Only for the 2ª order watersheds analyzed
4	BFO2	2nd order micro basins plus fragments
5	BO3	Only for 3rd order watersheds analyzed
6	BFO3	3rd order micro-watersheds plus fragments
7	BO4	Only for the 4th order watersheds analyzed
8	BFO4	4th order micro-watersheds plus fragments

Calculation of the morphometric variables of the watersheds

To calculate the variables, two maps had to be made: the altimetry map and the slope map. The altimetry map is actually the MDE cut out for the basin, which was assigned a color ramp and a layout was produced. The slope map was *input from* the basin's MDE. This product was generated in the Qgis *software* and when it generates the slope its product is in radian degrees, so using the *software*'s raster calculator we ran formula 1 on the slope raster (slope_raster).

$$tan(\text{"}slope\_raster\text{"} * 3.14159 / 180) * 100 \tag{1}$$

The result is the slope raster in percentages which has been reclassified by the Qgis *r.reclass* processing module following the classes defined by Embrapa (1979).

The variables considered were those that provided the basis for calculating the parameters: minimum (hmin) and maximum altitude in meters (hmax); average slope in % (Gbm); total length of the watercourses in each watershed of a given order in meters (Lt); length of the main river (Lp); vector distance in meters (dv); area in m^2 (A) and perimeter in meters (P).

The hmin, hmax and Gbm variables were processed in a GIS environment as follows: altimetry and slope information was collected for vector points in the centroid of each pixel of each thematic raster, by means of spatial consultation, indexing and field statistics of all the points, and thus the minimum, maximum and average altimetry and slope values were extracted for each micro-basin.

The variable Lt was obtained by adding up all the channels in each watershed. To define the main course of the river, the criterion of the longest stream in the watershed was taken into account, which is in accordance with Araùjo et al. (2007) and Christofoletti (1980). To calculate Lp, new *shape* files were made to represent the main rivers and calculate their lengths. The dv is defined by Salgado et al. (2009) as the distance in a straight line from the ends of the main river in the basin, and the length was calculated from the straight line in vector format. Finally, the area and perimeter were calculated using the field calculator in the attributes table.

It is worth noting that the procedure for obtaining all the variables was repeated for the data from each treatment.

Calculation of the morphometric parameters of the basins

Calculating the parameters will help to understand the processes of anthropogenic occupation and assist in planning the management of the basin (CARDOSO et al., 2006).

With the variables calculated for each watershed according to its treatment (Table 1), a database was created in Excel software to calculate the parameters and analyze Horton's laws.

First, the normality analysis of the micro-basins suggested by the parameters that govern Horton's laws was carried out, in order to validate whether the modeling satisfies the expected behavior of the basins. According to Christofoletti (1980), the parameters that underpin the laws of hydrographic basins are:

- **Bifurcation ratio** (R_b): this parameter is related to the law of the number of channels. It is defined as the ratio between the total number of segments of a certain order and the total number of segments of an immediately higher order. Using Strahler's ordering system, this result cannot be less than 2. The expression used for the calculation can be expressed as:

$$R_b = \frac{N_u}{N_{u+1}} \quad 2)$$

Where: N_u is the number of segments in a given order and N_{u+1} is the number of segments in the next higher order.

- **Ratio between average channel lengths** (RL_m): the length of channels can be represented by the letter L. If each order of **channels** is also represented by the letter u, the total sum of the channel lengths of each order will be L_u , and the total length of all the watercourses in a basin will be represented by L_t . To calculate the average length of river segments, L_m , divide the sum of the lengths of the channels of each order L_u by the number of segments found in the respective order N_u . In this way we use:

$$L_m = \frac{L_u}{N_u} \quad 3)$$

This relationship is the basis for the length law. channel parameterThe following equation is used

$$RL_m = \frac{Lm_u}{Lm_{u-1}} \quad 4)$$

Where: RL_m is the ratio between the average lengths of the channels; Lm_u is the average length of the channels in a given order, and Lm_{u-1} is the average length of the channels in the next lower order.

- **Channel gradient** (*Rgc*): the channel gradient is the ratio between the **maximum** difference in altitude between the point of origin and the end and the length of the respective river segment. Its purpose is to indicate the slope of watercourses, and it can be measured for the main river and for all segments of any order. This parameter is the basis for the slope law for canals, and can be calculated from the formula:

$$Rgc = \frac{Gmc_u}{Gmc_{u+1}} \quad 5)$$

Where: Gc_u is the average slope of channels of a certain order and Gc_{u+1} is the average slope of channels of an immediately higher order.

- **Ratio between basin areas** (R_a): as mentioned above, since each higher-order segment drains an increasingly larger area as the channel orders increase, the index seeks to relate the areas of the basins of subsequent orders, such as between the first and second order, between the second and third order, and so on.

By measuring the areas of the basins corresponding to each order, we obtain the absolute values for them; knowing the individual values and the number of cases, we can calculate the *average area* for the basins of a given order. The relationship between the areas of the basins is the basis for the law of the basin area of the canals, and can be expressed by the equation:

$$R_a = \frac{A_u}{A_{u+1}} \quad 6)$$

Where A_u is the average area of the basins for a given order and A_{u-1} is the average area of the basins of the next lower order.

These four parameters - formulas 2, 4, 5 and 6 - represent their respective laws, which are illustrated graphically by lines showing the correlation between the order of channels (abscissa axis) and the logarithm of the parameter value per channel order (ordinate axis). These lines correspond to the ideal behavior of river basins for Horton (1945) and Schumm (1956).

In the research, both the base variables and the parameters were calculated to be applied to the micro-basins to be generated, being applied according to the orders generated, so the analysis of the micro-basins will be repeated for each hierarchical order. Therefore, only the results of the parameters (F_f, K_c, I_c, D_d, *Is, Gb_m and dh*) were considered input data for the statistical analysis. The morphometric parameters of the basins were calculated according to: Cardoso et al. (2006); Horton (1945); Salgado et al. (2009); Teodoro et al. (2007); Tonello (2005) and Villela and Mattos (1975).

The **shape of the basin** (F_f) was obtained from the following formula:

$$F_f = \frac{A}{L_P{}^2} \qquad 7)$$

Where A is the area of the basin and L_p is the length of the main course of the basin.

The **compactness coefficient** (K_c) is obtained from the following formula:

$$K_c = \frac{(0{,}28\ .\ P)}{\sqrt{A}}$$

Where *P is the* perimeter of the basin and *A is* the area of the basin.

The Kc relates the shape of the basin to a circle, being the ratio between the perimeter of the basin and the circumference of a circle with an area equal to that of the basin. The closer it is to 1, the more circular the basin is.

The **Circularity index** (I_c), which indicates the shape of the basin, is obtained from the following formula:

$$I_c = \frac{(12{,}57\ .\ A)}{P^2} \qquad 9)$$

Where A is the area of the basin and P is the perimeter of the basin.

The order of the watercourse in turn is referred to according to the degree of branching of the basin (VILLELA; MATTOS, 1975). Thus, the order of the basin is based on its degree of branching; the higher the order of the basin, the more branched it is (HORTON, 1945).

Drainage **density** (Dd) is represented by the ratio between the total length reached by watercourses and the area of the basin (HORTON, 1945), which reflects the availability of channels for the linear flow of water and the degree of dissection of the relief resulting from the action of the drainage network, according to Salgado et al. (2009).

$$D_d = \frac{L_t}{A} \qquad 10)$$

Where L_t is the total length of the watercourses in the basin and A is the area of the basin.

In addition, the **Sinuosity Index** (I_s) refers to the ratio between the length of the main course of the basin and its vector length (distance in a straight line between the ends of the main course), which is a controlling factor of the speed of water flow (SALGADO et al., 2009), obtained by the following formula:

$$I_s = \frac{L_P}{d_v} \qquad 11)$$

Where L_p is the length of the main course of the basin and d_v is the vector distance.

The **average slope** (Gb_m) is calculated to assess the intensity of morphogenetic processes, including the dynamics of concentrated and diffuse (laminar) surface flows on the slopes. When associated with the maximum slope, it enables comparisons to be made of the maximum and average energy within watersheds (CHEREM, 2008). And the **altimetric amplitude** (dh) is the variation in the maximum and minimum amplitude of a basin (TEODORO et al., 2007).

Once the morphometric parameters had been calculated, the table with their values for each treatment in group 1 was indexed in the GIS environment. A core categorization was then assigned by means of quantiles (equal intervals) divided into 5 classes, for spatial representation of the parameters that most influenced the components analyzed (treatments in group 2).

Multivariate statistical analysis of drainage basins

After generating the morphometric parameters established above for the eight treatments, the database with the information was duly organized in the Micro Software Excel spreadsheet to generate principal component analysis (PCA) and cluster analysis (CA).

The PCA was carried out using BioEstat 5.0 software. The values of the morphometric parameters calculated in Excel were used as *input* data, from which the data was transformed using the base 10 log of each parameter. With the data transformed, the PCA was applied in the statistical program, resulting in: the correlation matrix between the parameters, the eigenvalue matrix (absolute and accumulated), the eigenvector matrix and the matrix containing the analysis value for each component in each watershed evaluated. - These values were used to show the variability of each component in each parameter in the respective sub-treatments considered in the research (Table 2).

The analysis will generally result in three new variables (PC) that most closely represent the variation in the database parameters, with each PC being strongly influenced by one or a few variables in the input set.

Table 2 - List of analysis sub-treatments

index	Treatment	Description
1	CP1BO1	Principal component 1 only for the 1ª order watersheds analyzed
2	CP1BFO1	Principal component 1 for 1ª watersheds plus fragments
3	CP2BO1	Principal component 2 only for the 1ª order watersheds analyzed
4	CP2BFO1	Principal component 2 for 1st order watersheds plus fragments
5	CP3BO1	Principal component 3 only for 1st order watersheds analyzed
6	CP3BFO1	Principal component 3 for 1st order micro-watersheds plus fragments
7	CP1BO2	Principal component 1 only for 2nd order watersheds analyzed
8	CP1BFO2	Principal component 1 for 2nd order watersheds plus fragments
9	CP2BO2	Principal component 2 only for 2nd order watersheds analyzed
10	CP2BFO2	Principal component 2 for 2nd order watersheds plus fragments
11	CP3BO2	Principal component 3 only for 2nd order watersheds analyzed
12	CP3BFO2	Principal component 3 for 2nd order watersheds plus fragments
13	CP1BO3	Principal component 1 only for 3rd order watersheds analyzed
14	CP1BFO3	Principal component 1 for 3rd order watersheds plus fragments
15	CP2BO3	Principal component 2 only for 3rd order watersheds analyzed

16	CP2BFO3	Principal component 2 for 3rd order watersheds plus fragments
17	CP3BO3	Principal component 3 only for 3rd order watersheds analyzed
18	CP3BFO3	Principal component 3 for 3rd order watersheds plus fragments
19	CP1BO4	Principal component 1 only for 4th order watersheds analyzed
20	CP1BFO4	Principal component 1 for 4th order micro-watersheds plus fragments
21	CP2BO4	Principal component 2 only for 4th order watersheds analyzed
22	CP2BFO4	Principal component 2 for 4th order micro-watersheds plus fragments
23	CP3BO4	Principal component 3 only for 4th order watersheds analyzed
24	CP3BFO4	Principal component 3 for 4th order micro-watersheds plus fragments

The purpose of defining these sub-treatments was to check which parameter most influences the variability of the data analyzed. Each principal component had a parameter with a higher eigenvector coefficient or weight, which means that it exerts a greater influence on the variability of the component in question when compared to the other eigenvectors.

To this end, the following conditional function was established for each sub-treatment: if the parameter is the one with the highest eigenvector weight, then a value of 1 is assigned, if the condition is not met, then a value of 0 is assigned. After this, all the values for each parameter were added up and it was thus possible to obtain a *ranking of* the parameters that concentrated the greatest variability in the data by means of the eigenvector coefficient in each component investigated.

With the components processed for each treatment, a cluster analysis was generated using the Action extension of MS Excel, using the non-hierarchical *K-means* method, which generated groups that varied according to the behavior of the data from each treatment, with the number of groups formed in the treatments varying. The result was a table in which the index code, which identifies each watershed, was allocated to the columns referring to the groups identified and, using this information, the table was related to the spatial features using the index column common to both files.

This cluster analysis was carried out in order to define management zones - where the characteristics of the site are similar, homogeneous - so that the territory of the micro-basin follows a territorial planning taking into account its geomorphological characteristics.

RESULTS AND DISCUSSION

Modeling the drainage network and the hierarchical micro-basins that make up the Apeù river basin from a hydrologically corrected digital elevation model (DEM)

The initial vector reference data for the channels was the 1:100,000 drainage base (Figure 4c) and matrix maps from the DSG database, which guided the digitization of the hydrography using RapidEye images (Figure 4b) and refined by Google Satellite (Figure 4a). It can be seen that the refinement corrected the course of the watercourse, which gave the MDE more reality when comparing Figure 4b and Figure 4a.

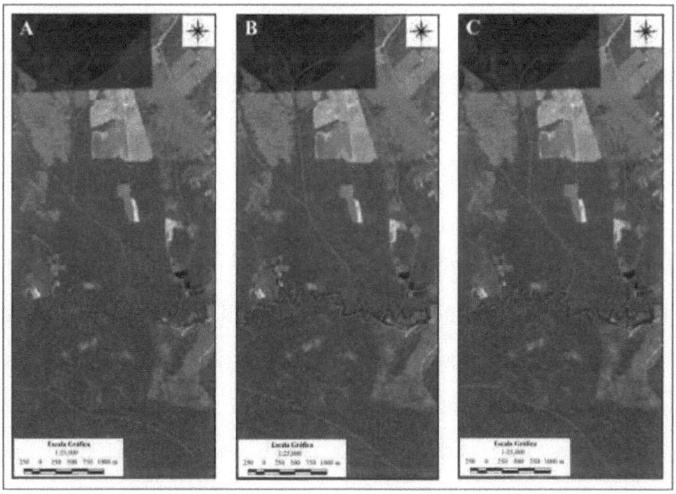

Figure 4 - Detail of the drainage for: the 1:100,000 DSG base (C); vectorized drainage with RapidEye images (B) and refined drainage with Google Satellite images (A).

Source: author

It can be seen that the process of hydrological correction of the MDE really succeeded in demarcating the vectorized drainage network (Figure 5b), so that it did not alter the general conformation of the original model (Figure 5a), except in the deepened part, even though the altimetry values were altered during the process.

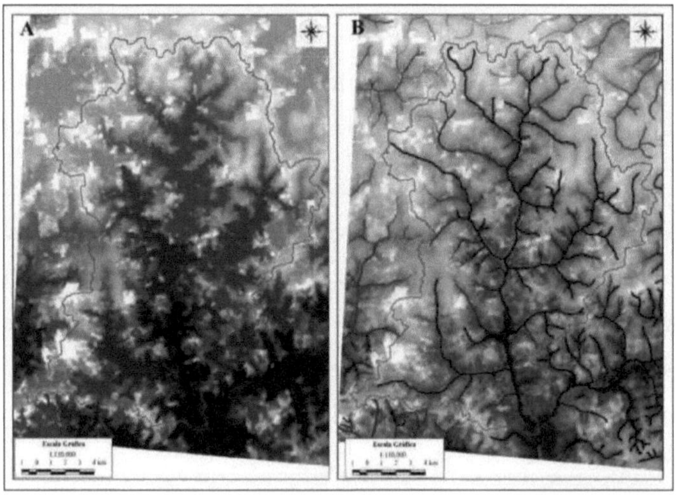

Figure 5 - Original pre-processed Topodata EAW (A) and hydrologically corrected Topodata EAW (B).

Source: author

Once the MDE had been run, it was processed to extract the drainage micro-basins and hydrographic network. When comparing the vectorized drainage (Figure 6a) and the automatically extracted drainage (Figure 6b), it is clear that the drainage deepening procedure preserved the vectorized channels, so the generated channel network is closer to reality, which generated well-modeled micro-basins. However, it was not possible to find a threshold that extracted the hydrography with 100% fidelity to the original vectorization. The threshold used was 300, which was the closest to the vectorization, so as not to eliminate large watercourses *that had already* been identified. With this threshold, some drainages were very small when compared to their micro-basins, but this was due to a scalar limitation of the data used; probably if the work had been applied to an MDE with a larger scale, the channels would have been demarcated more faithfully to their physiography on the ground.

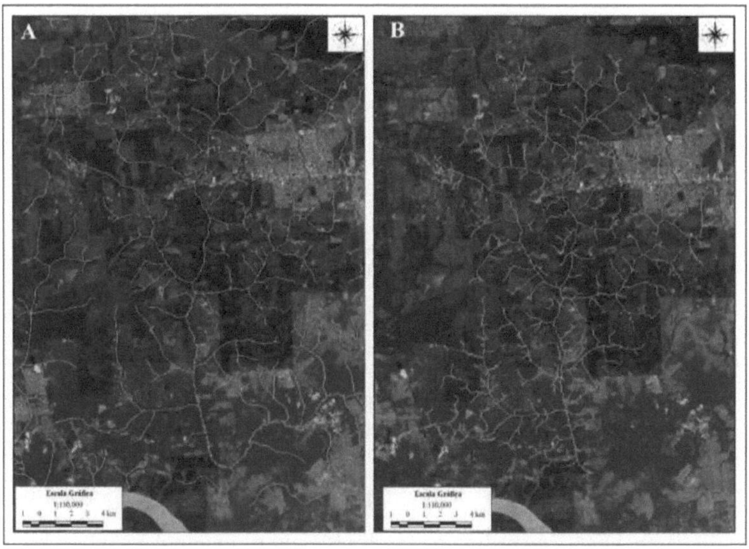

Figure 6 - Vectorized and refined drainage based on high-resolution images (A) and drainage extracted automatically by the basin modeling process in the TauDEM algorithm (B).

Source: author

The original Topodata DEM, with only the depressions corrected, was used to make the altimetry map (Figure 7), with the topographic amplitude of the BHRA being 61 meters and the slope map in percentages according to the Embrapa systematization (1979), so that the minimum slope identified was 0% (flat terrain), the average was 3% and the maximum was 21% (strongly undulating terrain) (Figure 8), with a more elongated shape (Ff equal to 0.31) and high runoff potential (Ic equal to 0.17).

The methodology of automatic detection of the watersheds allowed the study area to be individualized into 1ª , 2ª , 3ª , 4ª river order (Figure 9 and Figure 11) and into 1st, 2nd, 3rd, 4th river order plus fragments (Figure 10 and Figure 12), The 5th order identified was not included in the treatment due to the fact that it corresponds to the geographical limit of the basin, thus characterizing the BHRA as 5th order, so only the other orders generated enough observations (micro-basins) to continue the study. The number of modeled watersheds for each treatment is shown in Table 3.

Table 3 - Number of watersheds modeled per treatment analyzed.

index	Treatment	No. of micro-watersheds
1	BO1	305
2	BFO1	400
3	BO2	76
4	BFO2	174
5	BO3	14
6	BFO3	182
7	BO4	4
8	BFO4	89

The fragments representing unclassified drainage micro-basins are concentrated along the valleys, which according to Florenzano (2008) are prolonged depressions situated in the lower areas of the terrain that contain the highest order watercourses. These valleys can be clearly seen on the altimetry map (figure 7) where the talvegue passes and evidenced by the slope map which is predominantly flat. This is due to Strahler's process of fluvial hierarchization, in which the higher the fluvial order, the greater the length of the drainage channels, according to Horton's Law of channel length, which considers only the channels of the order analyzed, as can be seen in figures 9, 10, 11 and 12. It is also possible to see that as the orders increase, the shape pattern evolves from more circular shapes to more elongated and irregular shapes (figures 9, 10, 11 and 12).

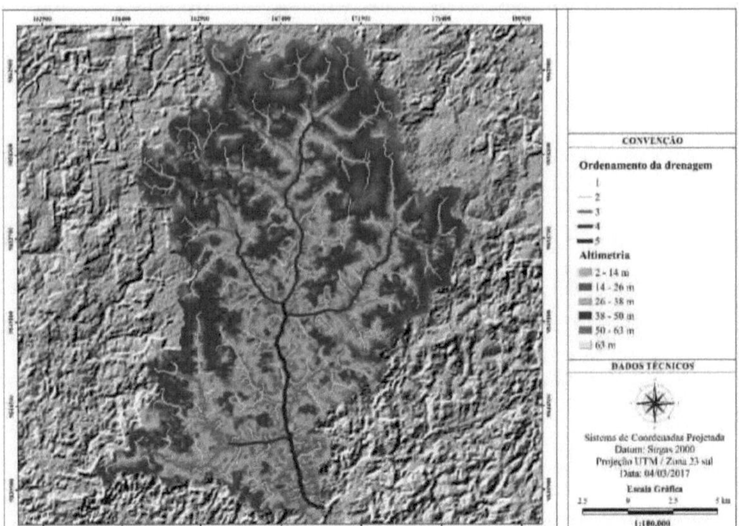

Figure 7 - Altimetry map obtained from the Topodata MDE, with depression correction made by the TauDEM

algorithm© 5.1.

Source: author

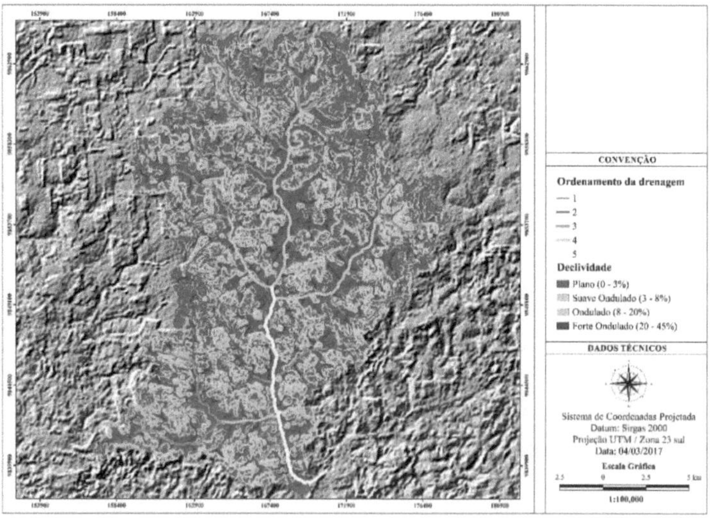

Figure 8 - Slope map classified according to the Embrapa systematization (1979).

Source: author

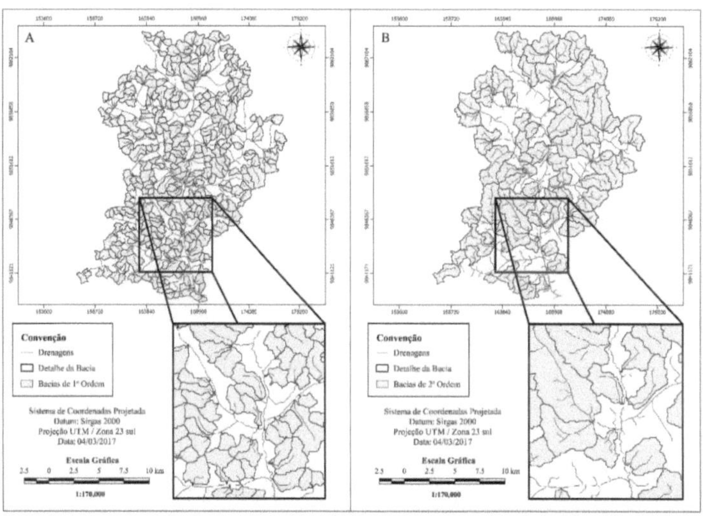

Figure 9 - 1ª order watersheds (A) and 2ª order watersheds (B) only for the orders analyzed.

Source: author

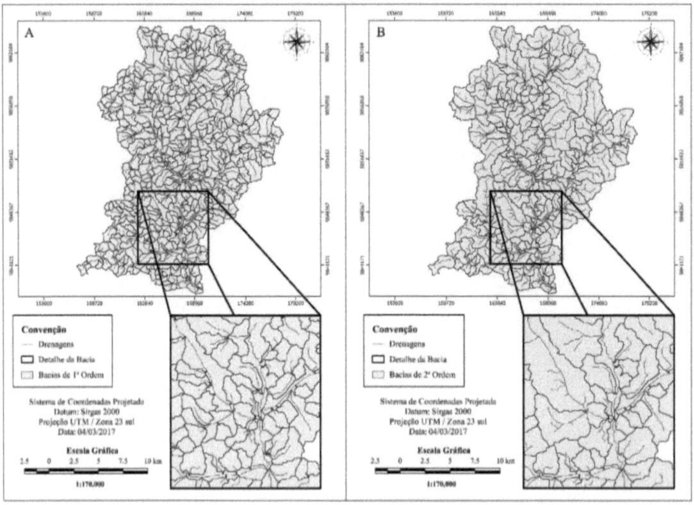

Figure 10 - 1ª order micro-watersheds (A) and 2ª order micro-watersheds (B) plus fragmentation of the entire study area.

Source: author

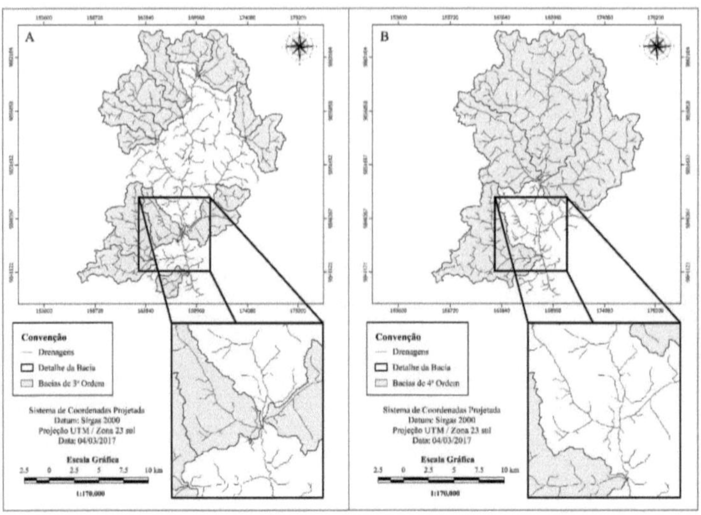

Figure 11 - 3ª order watersheds (A) and *4ª* order watersheds (B) only for the orders analyzed.

Source: author

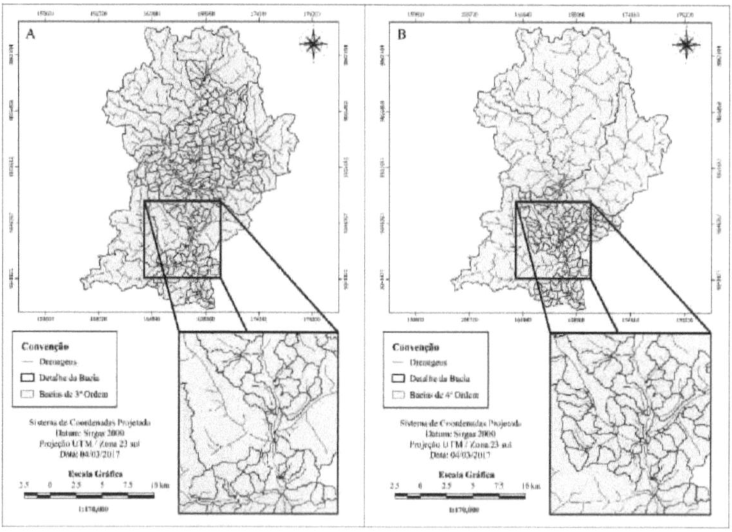

Figura 12 - 3ª order micro-watersheds (A) and 4ª order micro-watersheds (B) plus fragmentation of the entire study area.

Source: author

Table 4 shows the result of applying Horton's analysis, in order to show whether the methodology applied to delimit micro-basins and extract the hydrographic network follows the behavior expected by the author, in order to give more consistency to the data generated in the work.

Table 4 - Results of the application of Horton's Laws to the Apeù River MicroBasins. Where we have: number of segments of a given order (Nu), bifurcation ratio (Rb), average length of river segments of each order in *m* (Lm), ratio between average lengths of channels (RLm), average slope of channels of a given order in % (Gcu), gradient of channels (Rgc), average area of microbasins for a given order in *knr* (Au), ratio between areas of microbasins (Ra).

Order (u)	Naked	Log10 Nu	Rb	Lm (m)	Log10 Lm	RLm	Gcu	Log10 Gcu	Rgc	Au (km)^2	Log10 Au	Ra
1ª	305	2.48	4.01	578.37	2.76	-	1.11	0.04	0.71	0.61	- 0.22	-
2ª	76	1.88	5.43	1,437.21	3.16	2.48	1.55	0.19	0.67	2.75	0.44	4.52
3a	14	1.15	3.50	3,728.92	3.57	2.59	2.31	0.36	2.23	13.10	1.12	4.76
4a	4	0.60	4.00	7,775.80	3.89	2.09	1.03	0.01	0.90	64.83	1.81	4.95
5a	1	-	-	14,638.90	4.17	1.88	1.14	0.06	-	316.04	2.50	4.88
Average			4.24			2.26			1.27			4.78

Table 5 presents a summary of the behavior of a watershed expected by Horton and Figure 13 shows the results found in this work.

Table 5 - Summary of Horton's expected micro-watershed behavior (CHRISTOFOLETTI, 1980)

Horton's expected behavior
Law of No. channels (Nu): inverse geometric series
Channel length law (Lm): direct geometric series
Law of channel slope (Gcu): inverse geometric series
Law of the area of the micro-basin of canals (Au): direct geometric series

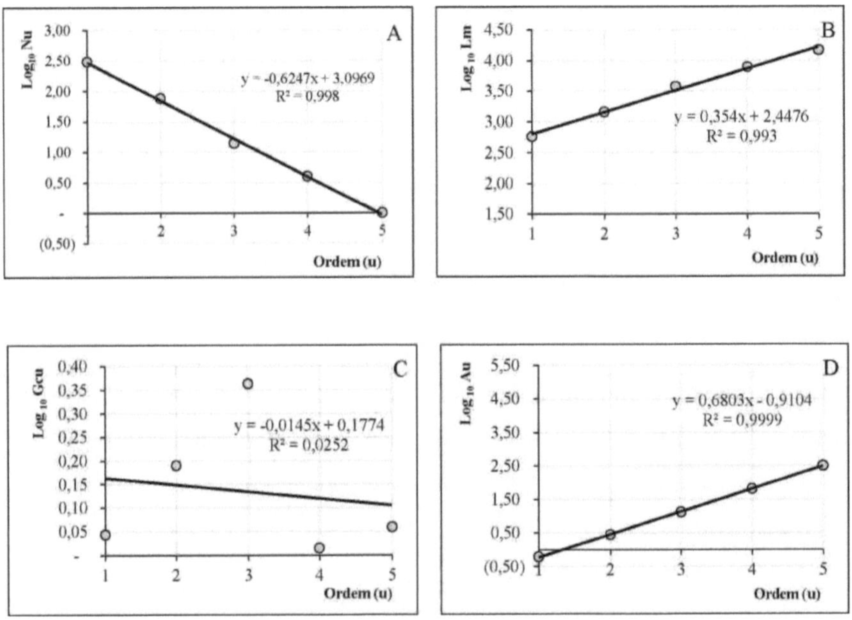

Figura 13 - Horton's Laws represented graphically. Where we have: A - number of segments of a given order (Nu), B - average length of river segments of each order in *m* (Lm), C - average slope of channels of a given order in *%* (Gcu), D - average basin area for a given order in km^2 (Au).

Source: author

Figure 13 shows that all the variables, except Gcu, met the expected behavior - represented by the regression line, where R^2 is the coefficient of determination that shows how well the mathematical model predicts the analyzed variable, the closer it is to 1 the better the model fits the behavior of the variable. Gcu's behavior was not significantly represented by the linear model, due to the fact that the slope measurements along the channel were sampled in the non-hydrologically corrected slope model, because as was pointed out in the methodology of the work, the corrected MDE was only used to delimit the micro-basins and automatically extract the drainage. Therefore, when using a hydrographic network that is not faithful to the slope model from which the values were sampled, the watershed line does not follow the lower slopes of the terrain.

As a result, Gcu was calculated for the drainage extracted from the original model and the result of this procedure is shown in Figure 14, which shows what Horton expected (inverse geometric series), confirming that the low value of R^2 for Gcu in Figure 13c is due to the change in the MDE, since the linear model in Figure 14 represents the real behavior of the variable.

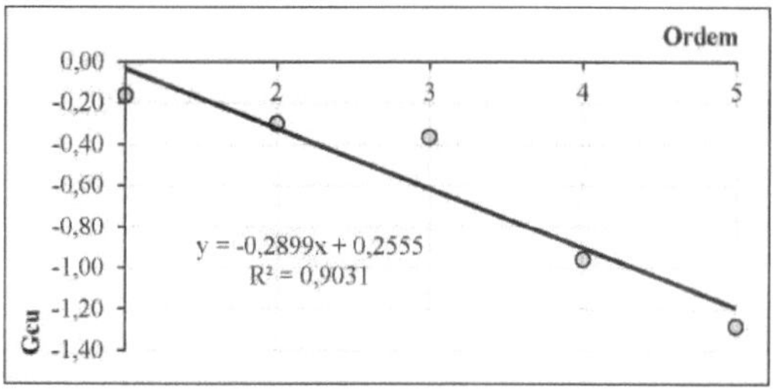

Figura 14 - Horton's law of the average slope of channels of a given order in % (Gcu) represented graphically, for the MDE model without hydrological correction.

Source: author

The results found in this study are in line with Vestena, Checchia and Kobiyama (2006). In that study, the authors did not use a corrected model but a 1:50,000 topographic map containing topographic data to generate the terrain model and model the Caeté watershed, and the results of the behavior of the prediction model and R2 were similar. This reinforces the methodology used in this research, both for correcting the MDE and for modeling.

Principal component analysis for morphometric characterization of drainage basins

The PCA was applied to the following morphometric parameters: shape factor (Ff), compactness coefficient (Kc), circularity index (Ic), drainage density (Dd), sinuosity index (Is), average slope of each watershed (Gbm) and altimetric amplitude (dh), and the analysis was applied to the data set of these parameters for each of the drainage watershed treatments.

The PCA calculated seven main components, with the first three components concentrating around 80% of the total variance of the input data (morphometric parameters) and the seventh component having zero variance for all the observations (micro-watersheds), as it had low representativeness of the data variance. In order to show the result of the variability of the components, it was decided to analyze only the first three components, due to their representativeness in the variability of the data. The results for data variance are shown in Figure 15 and Figure 16.

The result of applying PCA provided a hierarchy of the morphometric parameters analyzed, taking into account the coefficient of the eigenvector with the greatest weight, so that this was the variable that most influenced each of the subtreatments. Thus, using a conditional function, it was possible to identify the parameters that most influenced the variability of the data.

According to Florenzano (2008), morphometry refers to the quantitative aspects of the relief. Thus, following the Embrapa (1979) classification, the relief of the Apeù river basin was characterized as flat, gently undulating, undulating and strongly undulating, as shown in Figure 8. Also according to Florenzano (2008), the quantitative relief data allows for an association with the forms of relief: flat relief is associated with plains, terraces, tablelands and plateaus; gently undulating relief is associated with hills; undulating relief is associated with hills and hillslopes; strongly undulating relief is associated with hills and hillslopes; mountainous relief is associated with mountains and hillslopes; and steep relief is associated with mountains and hillslopes.

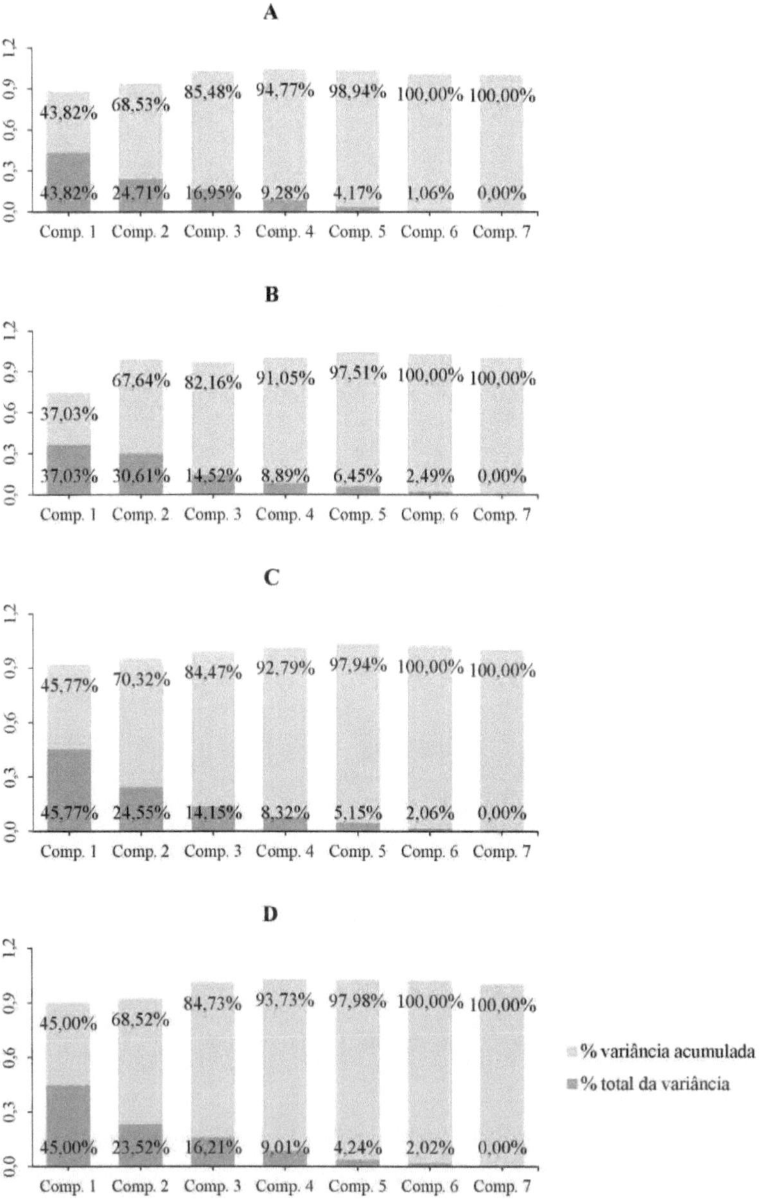

Figura 15 - Relative importance of the principal components for: (A) 1ª order micro-watersheds plus fragments; (B) 2ª order micro-watersheds plus fragments; (C) 3ª order micro-watersheds plus fragments and (D) 4a order micro-watersheds plus fragments.

Source: author

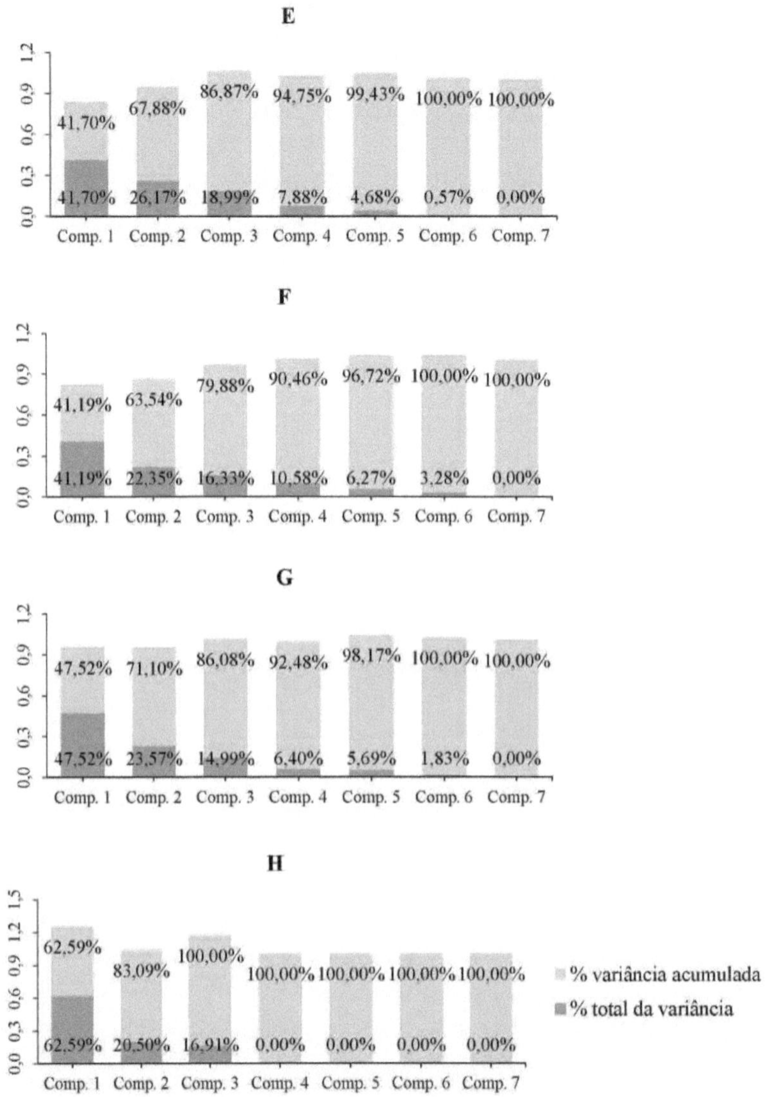

Figure 16 - Relative importance of the principal components for: (E) 1ª order watersheds; (F) 2ª order watersheds; (G) 3ª order watersheds and (H) 4ª order watersheds.

Source: author

Analysis of the 1st order watersheds ª

As shown in Table 6, it can be seen that of the parameters analyzed and used as *input* for the PCA, all are correlated in some way, so that there was no nullity between the variables analyzed for all the treatments analyzed. This result provides support for validating the component analysis, since (Vicini; Souza (2005)) states that there must be a correlation structure between the variables, otherwise the information generated from the PCA cannot be used. According to the same author, zero correlation coefficients between synthetic variables are due to the variables being independent of each other.

Table 6 also shows that among the parameters analyzed, those with the highest correlation were: Kc and Ic; Ff and Dd, which proved to be inversely proportional.

Table 6 - Correlation matrix of morphometric parameters and principal components, only for the P-order watersheds analyzed. Only the first three principal components, which accumulated around 80% of the total variance of the data, were considered for analysis.

	Ff	**Kc**	**Ic**	**Dd**	**Is**	**Gbm**	**dh**	**CP 1**	**CP 2**	**CP 3**
Ff	**1.00**	-0.27	0.27	-0.95	**-0.56**	-0.13	-0.31	0.87	0.21	0.34
Kc	-0.27	**1.00**	-1.00	0.33	0.10	-0.06	0.04	**-0.63**	0.68	0.36
Ic	0.27	-1.00	**1.00**	-0.33	-0.10	0.06	-0.04	**0.63**	**-0.68**	-0.36
Dd	-0.95	0.33	-0.33	**1.00**	**0.51**	0.15	0.26	**-0.88**	-0.15	-0.29
Is	**-0.56**	0.10	-0.10	**0.51**	**1.00**	0.04	0.16	**-0.59**	-0.22	**-0.48**
Gbm	-0.13	-0.06	0.06	0.15	0.04	**1.00**	**0.66**	-0.23	**-0.65**	0.60
dh	-0.31	0.04	-0.04	0.26	0.16	**0.66**	**1.00**	**-0.42**	**-0.61**	**0.52**
CP 1		**-0.63**	**0.63**	**-0.88**	**-0.59**	-0.23	**-0.42**	**1.00**	0.00	0.00
CP 2	0.21		**-0.68**	-0.15	-0.22	**-0.65**	**-0.61**	0.00	**1.00**	0.00
CP 3	0.34	0.36	-0.36	-0.29	**-0.48**		**0.52**	0.00	0.00	**1.00**

The correlation between Kc and Ic was 100% (Table 6), as both are related to the circular shape of the watershed. The higher the Kc value, the lower the risk of flooding, while for Ic the lower the value, the lower the risk of flooding, thus observing that the values have inversely proportional magnitudes that point to the same result. In addition, the characteristic linked to them is inversely proportional, i.e. the greater the surface runoff, the lower the risk of flooding, since the water load is more easily drained. It is important to note that both Kc - with higher values, distant from 1 - and Ic - with lower values - reflect a more irregular and elongated watershed, as mentioned by Schmitt and

Moreira (2015) and evidenced in the reference values of the morphometric parameters shown in Table 7.

Table 7 - Reference values for the morphometric parameters analyzed

Parameter	Reference	Un.	Meaning	Source
Coefficient of compactness (Kc)			Risk of flooding in the basin	
	1.00 - 1.25	ad.*	High	Schmitt and Moreira (2015)
	1.25 - 1.50	ad.	Medium	
	>1.50	ad.	Bass	
Form factor (Ff)	1.00 - 0.75	ad.	High	Santos, Silva and Braz, (2015)
	0.75 - 0.50	ad.	Medium	
	<0.50	ad.	Bass	
index of circularity (Ic)			Surface runoff from the basin	
	> 0.51	ad.	Bass	Schmitt and Moreira (2015)
	0.51	ad.	Medium	
	< 0.51	ad.	High	
Density of drainage (Dd)			Basin water availability	
	< 0.5	km/km^2	Low	Santos et al. (2012)
	0.50 - 2.00	km/km^2	Average	
	2.00 - 3.50	km/km2	High	
	> 3.50	km/km2	Very high	
sinuosity index (Is)			Channel shape	
	1.00	ad.	Retilineo	(Teodoro et al. (2007)
	1.00 - 2.00	ad.	Transitional (regular and irregular)	
	>2.00	ad.	Tortuous	

Note: * refers to dimensionless values.

Villela and Mattos (1975), mention that the parameters that describe the shape of the basin, trying to associate them with known geometric shapes, such as Kc for the circular shape and Ff for the rectangle shape, are dimensionless numbers that do not depend on the size of the basin, so that a Kc equal to 1 corresponds to a circular basin, and the more irregular the basin, the higher the compactness index, while for the Ff, the lower its value, the lower the possibility of flooding, and the higher its value, the greater the probability of flooding in the basin. Also according to the author, the more circular the basin, the greater the possibility of heavy rainfall simultaneously covering the entire length of the basin, so the contribution of the tributaries reaches the main watercourse at various points along the talvegue, thus concentrating all the water load at a single point.

Considering the above, it is understandable why Ff did not correlate significantly (table 6) with the other two shape characterization parameters, Kc and Ic, since Ff is similar to rectangular shapes, and Kc and Ic to circular shapes.

According to Christifoletti (1980), drainage density is related to water runoff and the total length of watercourses present in the basin area, as well as reflecting the rate of water infiltration into the soil. Consequently, the correlation between Ff and Dd is strong and inversely proportional (table 6) because the higher the form factor (higher flood risk), the lower the drainage density, and the lower the runoff potential, the greater the risk of water accumulating in the basin and flooding due to rainfall.

With the results of the eigenvector coefficients from the PCA, it was possible to observe that the parameters that showed the highest correlation with the principal components analyzed (Table 6) were those that had the greatest influence on the variability of their respective component (Table 8), such as the parameters Ff, Kc and Gbm, which in addition to having the highest correlation, also had the greatest weight for the 1st, 2nd and 3rd components, respectively. These results refer to basins without fragmentation.

Table 8 - Matrix of PCA eigenvector coefficients. Morphometric parameters that had the greatest weight in the variance of the data within each principal component, only for the 1ª order watersheds analyzed. Only the first three principal components, which accumulated around 80% of the total variance in the data, were considered for analysis.

	Ff	Kc	Ic	Dd	Is	Gbm	dh
CP 1	0.51	-0.37	0.37	-0.52	-0.35	-0.14	-0.25
CP 2	0.15	0.50	-0.50	-0.11	-0.16	-0.48	-0.45
CP 3	0.29	0.32	-0.32	-0.25	-0.41		0.45

For the micro-basins with fragments, Table 9 shows the entire correlation between the variables Kc and Ic, as observed for the micro-basins without fragments (Table 6). The parameters Ff, dh and Kc had the highest correlation, as well as the greatest weight for the 1st, 2nd and 3rd components, respectively (Table 6 and Table 10).

Table 9 - Correlation matrix of morphometric parameters and principal components for the 1ª order watersheds plus fragments. Only the first three principal components, which accumulated around 80% of the total variance of the data, were considered for analysis.

	Ff	Kc	Ic	Dd	Is	Gbm	dh	CP 1	CP 2	CP 3
Ff	**1.00**	**-0.54**	**0.54**	**-0.85**	**-0.49**	0.04	-0.21	0.89	-0.01	0.34
Kc	**-0.54**	**1.00**	-1.00	0.37	0.19	-0.12	0.19	**-0.84**	0.01	0.52
Ic	**0.54**	-1.00	**1.00**	-0.37	-0.19	0.12	-0.19	**0.84**	-0.01	**-0.52**
Dd	**-0.85**	0.37	-0.37	**1.00**	0.33	-0.20	-0.11	**-0.74**	-0.29	-0.43
Is	**-0.49**	0.19	-0.19	0.33	**1.00**	0.06	0.18	**-0.50**	0.17	**-0.57**
Gbm	0.04	-0.12	0.12	-0.20	0.06	**1.00**	**0.66**	0.11	**0.84**	-0.11
dh	-0.21	0.19	-0.19	-0.11	0.18	**0.66**	**1.00**	-0.22	0.90	0.06
CP 1	0.89	**-0.84**	**0.84**	**-0.74**	**-0.50**	0.11	-0.22	**1.00**	0.00	0.00
CP 2	-0.01	0.01	-0.01	-0.29	0.17	**0.84**	0.90	0.00	**1.00**	0.00
CP 3	0.34		**-0.52**	**-0.43**	**-0.57**	-0.11	0.06	0.00	0.00	**1.00**

Table 10 - Matrix of PCA eigenvector coefficients. Morphometric parameters that had the greatest weight in the variance of the data within each component for the 1ª order micro-basins plus fragments. Only the first three principal components, which accumulated around 80% of the total variance in the data, were considered for analysis.

	Ff	Kc	Ic	Dd	Is	Gbm	dh
CP 1	0.51	-0.48	0.48	-0.42	-0.28	0.06	-0.13
CP 2	-0.01	0.00	0.00	-0.22	0.13	0.68	0.68
CP 3	0.31		-0.48	-0.40	-0.53	-0.08	0.05

To better visualize the values of the components and parameters, high values were identified by the colors blue and green; intermediate values by yellow and low values by orange and red.

By comparing the spatial distribution of the 1ª component (Figure 17) and the respective eigenvectors with the highest weight (Figure 18), it can be seen that there is a distribution pattern that follows the distribution of the main component under analysis, both for the analysis with and without fragments. However, in the spatial arrangement with fragmentation (Figure 17b and Figure 18b) it is possible to see a pattern of distribution and grouping of the values that is more perceptible than in the distributions only for the orders analyzed (Figure 17a and Figure 18a). This is because the fragments fill in the lack of information observed for the microwatersheds of the order analyzed (without fragments).

Considering the reference values in Table 7, a distribution pattern can be seen, so that the highest Ff values (Figure 18) for the micro-watersheds have a more circular shape, which is consistent with a greater risk of flooding, while the highest values have a more elongated and irregular shape, being less susceptible to flooding, and intermediate values have a shape divided between elongated and circular micro-watersheds. A similar pattern of spatial representation of the micro-watersheds can be observed for the 1st component of the micro-watersheds under analysis (Figure 17).

Therefore, the PCA for sub-treatments 1 and 2 made it possible to group the Ff parameter more efficiently.

Analyzing the spatial distribution of the eigenvectors of the Kc parameter, which was the one with the greatest weight in the 2nd component of the micro-basins without fragments (Figure 20a), it is clear that the micro-basins are in a medium to low flood risk pattern, with some micro-basins with the lowest values shown - in red - tending towards the circular shape of the micro-basin, since their values are close to 1. When compared with the distribution of the 2nd component of the micro-watersheds without fragmentation (Figure 19a), it can be seen that the PCA de-characterizes the distribution pattern of the parameter that most influences it, so that it includes more elongated micro-watersheds for the lowest values of the component in question - shown in red.

Similarly, the distribution pattern of the 2^{a} component with fragmentation (Figure 19b) does not follow the same spatial arrangement as the dh data (Figure 20b).

For the treatments shown in Figure 19 and Figure 20, despite the correlation between the parameter and the variable, when presented spatially, they did not maintain a distribution pattern common to both.

The spatialization of the values of Gbm, the eigenvector with the greatest weight for the 3rd component of the basins without fragmentation (Figure 22a), shows a certain pattern of similarity to the spatial distribution of the component in question (Figure 21a).

The spatial representation of Kc in Figure 22b is consistent with what was expected, with lower values showing circular micro-watersheds (more prone to flooding) and the other intermediate to high values tending towards an increasingly elongated shape. According to the classification in Table 7, the flood risks that occur for the treatment in Figure 22b are medium and low.

When comparing Figure 22b with Figure 21b, it can be seen that the 3rd component of the micro-basins with fragments groups together more elongated and irregular basins that did not exist in the spatialization of the Kc parameter.

For the three components analyzed in the 1st order, the micro-basins with fragmentation made it possible to see the regionalization of both the components and the parameters with the highest eigenvector coefficients more clearly. When comparing the three main components with their respective eigenvectors, which had the greatest influence on their variability, it is easier to see this relationship in the 1st component when spatializing them, with Ff having the greatest influence, followed by Ic for the micro-basins with fragments (Table 10). This may be due to the fact that the 1st component concentrates the greatest variability in the data and thus produces a clearer spatial arrangement, consistent with the values of the parameters that most influence the component.

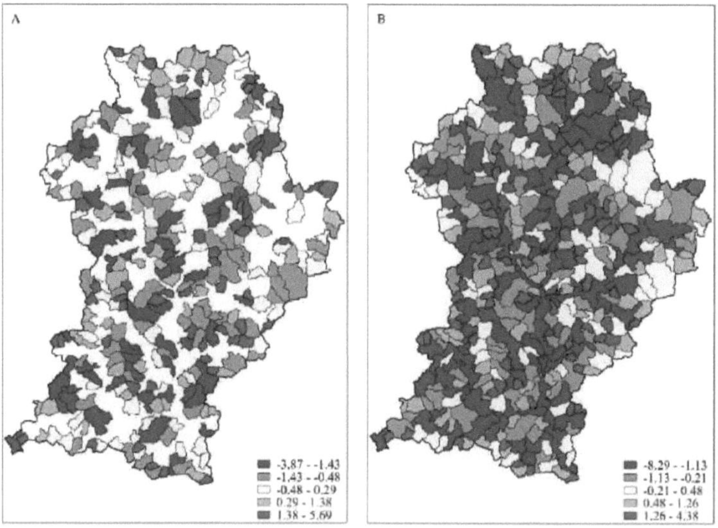

Figura 17 - ACP for the 1ª component of the 1ª order watersheds: only for the order analyzed (A) and with fragmentation (B).

Source: author

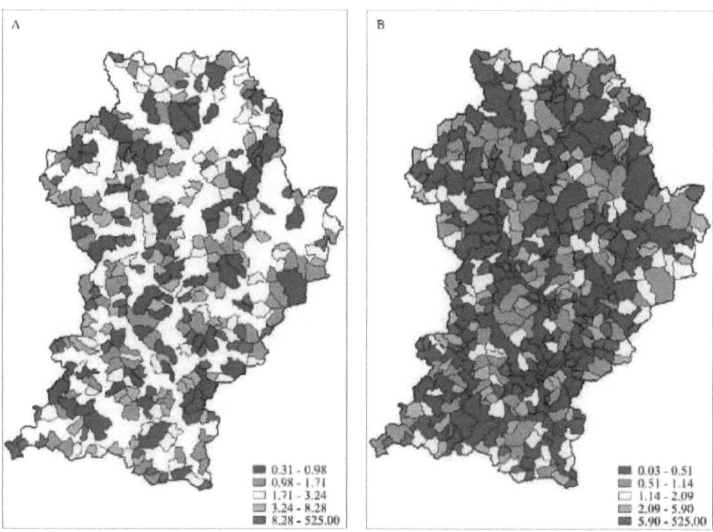

Figura 18 - Parameters representing the maximum eigenvector coefficient in the 1ª component of the P-order watersheds: Ff only for the order analyzed (A) and Ff with fragmentation (B).

Source: author

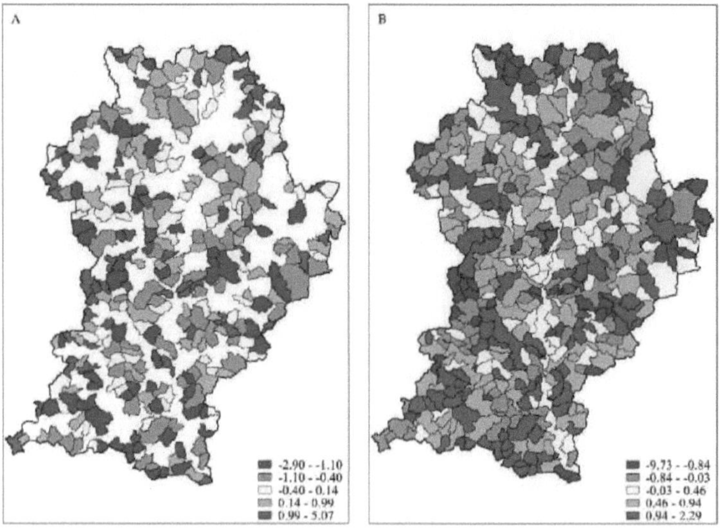

Figura 19 - ACP for the 2ª component of the 1ª order watersheds: only for the order analyzed (A) and with fragmentation (B).

Source: author

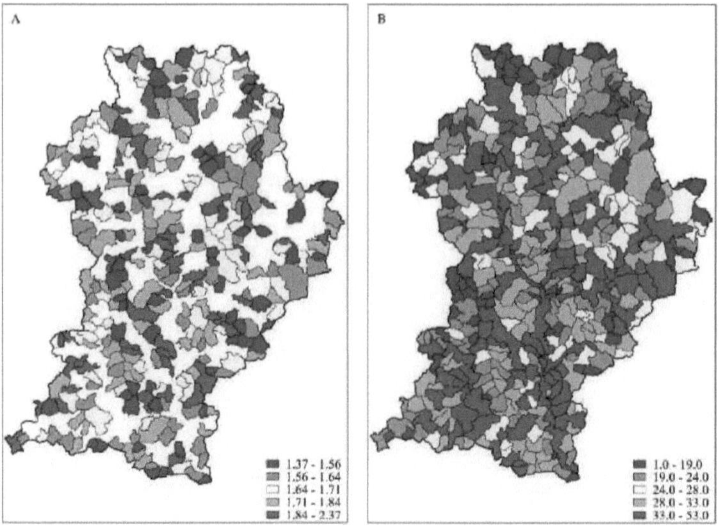

Figura 20 - Parameters that represented the maximum eigenvector coefficient in the 2ª component of the P-order watersheds: Kc only for the order analyzed (A) and dh with fragmentation (B).

Source: author

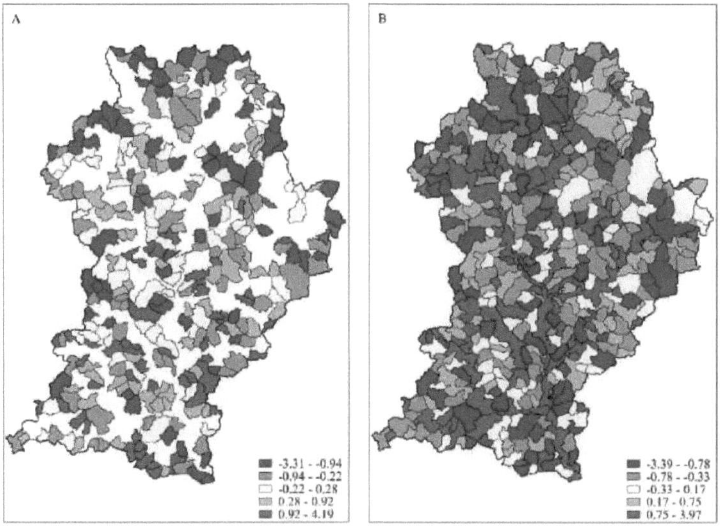

Figura 21 - PCA for the 3rd component of the 1ª order watersheds: only for the analyzed order (A) and with fragmentation (B).

Source: author

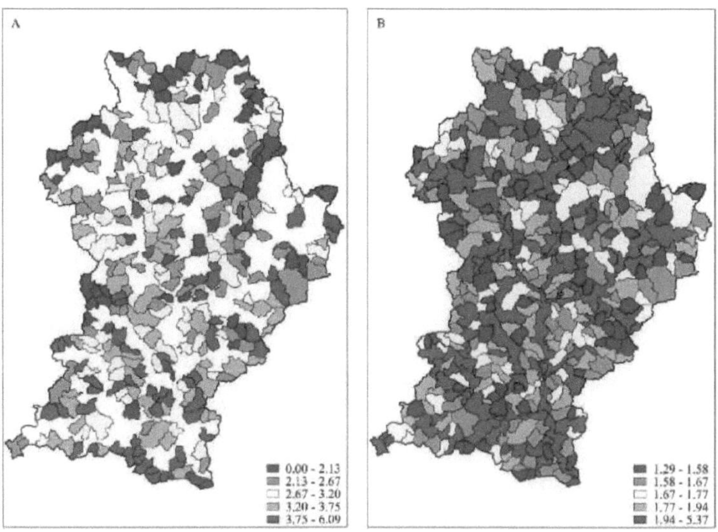

Figura 22 - Parameters that represented the maximum eigenvector coefficient in the 3ª component of the 1ª order watersheds: Gbm only for the order analyzed (A) and Kc with fragmentation (B).

Source: author

Analysis of the 2ª order watersheds

ªTables 11 and 13 show all the correlations between the parameters and between them and the components, so that there was no nullity between the input parameters for the PCA, making it feasible to use it for analysis. In the same tables, it can be seen that Kc and Ic are fully correlated, as observed for the first-order watersheds (tables 6 and 9).

The parameters Ic and Gbm were the ones that most influenced the variance of the 1st and 2nd components (table 12), respectively, which had the highest negative correlations (table 11) for the aforementioned components. However, the coefficient with the highest eigenvector (Kc) for the 3rd component (table 12) was not the parameter with the highest correlation with this component (table 11).

The parameters that most influenced the 1st, 2nd and 3rd components (table 14) were Ff, Ic and Gbm, respectively, and these were the ones with the highest correlations within each of these components (table 13).

Table 11 - Correlation matrix of morphometric parameters and principal components, only for the 2nd order watersheds analyzed. Only the first three principal components, which accumulated around 80% of the total variance of the data, were considered for analysis.

	Ff	Kc	Ic	Dd	Is	Gbm	dh	CP 1	CP 2	CP 3
Ff	**1.00**	**-0.61**	**0.61**	-0.37	-0.42	0.20	-0.17	**-0.83**	0.01	-0.03
Kc	**-0.61**	**1.00**	-1.00	0.24	0.25	-0.13	0.06	**0.90**	0.08	-0.36
Ic	**0.61**	-1.00	**1.00**	-0.24	-0.25	0.13	-0.06	-0.90	-0.08	0.36
Dd	-0.37	0.24	-0.24	**1.00**	0.42	0.29	0.04	**0.50**	-0.43	**0.59**
Is	-0.42	0.25	-0.25	0.42	**1.00**	-0.16	-0.01	**0.55**	0.02	0.65
Gbm	0.20	-0.13	0.13	0.29	-0.16	**1.00**	**0.47**	-0.16	-0.89	-0.05
dh	-0.17	0.06	-0.06	0.04	-0.01	**0.47**	**1.00**	0.10	**-0.76**	-0.33
CP 1	**-0.83**	**0.90**	-0.90	**0.50**	**0.55**	-0.16	0.10	**1.00**	0.00	0.00
CP 2	0.01	0.08	-0.08	-0.43	0.02	-0.89	**-0.76**	0.00	**1.00**	0.00
CP 3	-0.03	-0.36	0.36	**0.59**		-0.05	-0.33	0.00	0.00	**1.00**

Table 12 - Matrix of PCA eigenvector coefficients. Morphometric parameters that had the greatest weight in the variance of the data within each principal component, only for the 2nd order watersheds analyzed. Only the first three principal components, which accumulated around 80% of the total variance in the data, were considered for analysis.

	Ff	**Kc**	**Ic**	**Dd**	**Is**	**Gbm**	**dh**
CP 1	0.49	-0.53	0.53	-0.29	-0.32	0.10	-0.06
CP 2	-0.01	-0.06	0.06	0.34	-0.01	0.71	0.61
CP 3	0.03		-0.34	-0.55	-0.61	0.05	0.31

Table 13 - Correlation matrix of morphometric parameters and principal components for the 2ª order watersheds plus fragments. Only the first three principal components, which accumulated around 80% of the total variance of the data, were considered for analysis.

	Ff	**Kc**	**Ic**	**Dd**	**Is**	**Gbm**	**dh**	**CP 1**	**CP 2**	**CP 3**
Ff	**1.00**	-0.28	0.28	**-0.54**	**-0.53**	-0.25	**-0.54**	0.84	-0.25	0.25
Kc	-0.28	**1.00**	-1.00	-0.29	0.15	-0.08	0.21	**-0.59**	**-0.78**	0.05
Ic	0.28	-1.00	**1.00**	0.29	-0.15	0.08	-0.21	**0.59**	0.78	-0.05
Dd	**-0.54**	-0.29	0.29	**1.00**	0.27	0.36	0.17	-0.35	**0.74**	-0.24
Is_	**-0.53**	0.15	-0.15	0.27	**1.00**	0.08	0.30	**-0.61**	0.17	**-0.54**
Gbm	-0.25	-0.08	0.08	0.36	0.08	**1.00**	**0.50**	-0.37	**0.49**	0.64
dh	**-0.54**	0.21	-0.21	0.17	0.30	**0.50**	**1.00**	**-0.72**	0.20	**0.42**
CP 1	0.84	**-0.59**	**0.59**	-0.35	**-0.61**	-0.37	**-0.72**	**1.00**	0.00	0.00
CP 2	-0.25	**-0.78**	0.78	**0.74**	0.17	**0.49**	0.20	0.00	**1.00**	0.00
CP 3	0.25	0.05	-0.05	-0.24	**-0.54**		**0.42**	0.00	0.00	**1.00**

Table 14 - Matrix of PCA eigenvector coefficients. Morphometric parameters that had the greatest weight in the variance of the data within each component for the 2ª order micro basins plus fragments. Only the first three principal components, which accumulated around 80% of the total variance of the data, were considered for analysis.

	Ff	**Kc**	**Ic**	**Dd**	**Is**	**Gbm**	**dh**
CP 1	0.52	-0.37	0.37	-0.22	-0.38	-0.25	-0.45
CP 2	-0.17	-0.53	0.53	0.51	0.12	0.33	0.14
CP 3	0.25	0.05	-0.05	-0.24	-0.54		0.42

As the correlation between Ic and CP 1 was strong and inversely proportional (table 11) for the micro-watersheds without fragments, so the highest values for the 1st component (figure 23a) showed the most circular micro-watersheds, the same pattern of shape can be observed for the Ic parameter (figure 24a), so that all the values for Ic were considered low, according to the classification shown in table 7, in relation to the micro-watershed's runoff potential. Furthermore, the spatialization of the 1st component without fragmentation (figure 23a) follows the similar pattern found for the spatial arrangement of the eigenvector that most influences it (figure 24a).

When we look at the spatialization of the 1st component for the micro-watersheds with fragmentation (figure 23b), the higher values show more circular micro-watersheds and the lower values show more elongated micro-watersheds, so they are arranged more efficiently than in the representation of the micro-watersheds made only for the parameter that most represented the variance of the 1st component, Ff (figure 24b).

The spatial representation of the Gbm parameter (figure 26a), which represents more of the variance of the 2nd component for the micro-watersheds without a fragment, shows little similarity in spatial pattern when compared to its respective component (figure 25a).

The Ic values for the micro-watersheds with fragmentation (figure 26b) are better represented, considering an evolutionary pattern of the shape from circular to more elongated and irregular, than by the spatial arrangement shown in the 2nd component with fragmentation (figure 25b).

It can be seen that the Kc values for the micro-watersheds without fragmentation for the order under analysis (figure 28a) are classified as only micro-watersheds with medium and low flood risk, but there are still more circular micro-watersheds for the lowest values. And when we look at the reference component, the 3rd component without fragmentation (figure 27a), it arranges the watersheds to include the more elongated ones, but still does not maintain the pattern of similarity with the spatialization of the parameter that most influences it, Kc.

The spatial representation of the Gbm parameter (figure 28b), which represents more of the variance of the 3rd component for the micro-watersheds with fragments, shows little similarity in spatial pattern when compared to its respective component (figure 27b).

As observed for the analysis of the 1st[a] order, the micro-basins with fragmentation made it possible to see the regionalization of both the components and the parameters with the highest eigenvector coefficients more clearly. When comparing the three main components with their respective eigenvectors, which had the greatest influence on their variability, it is easier to see this relationship in the 1st component when spatializing them, and for the 2ndª order, the Ff had the greatest influence, followed by the Ic for micro basins with fragments (table 14).

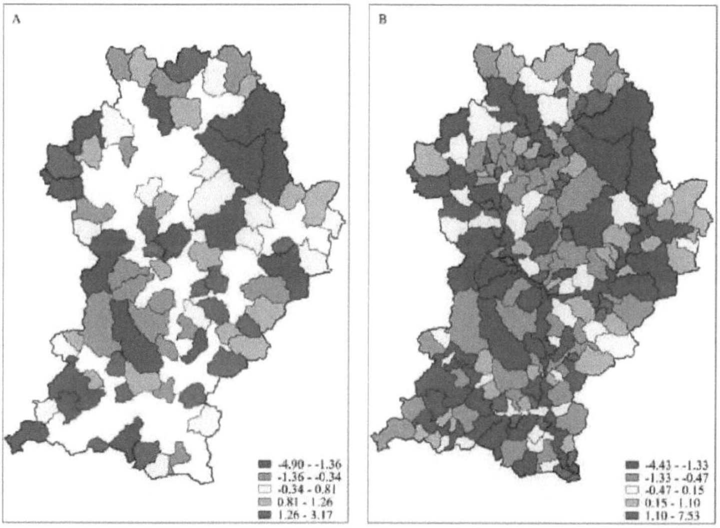

Figure 23 - PCA for the P component of the 2nd order watersheds: only for the order analyzed (A) and with fragmentation (B).

Source: author

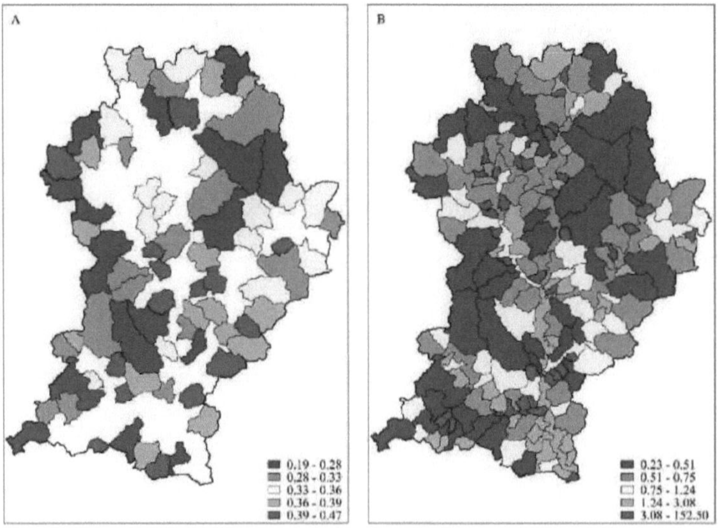

Figura 24 - Parameters that represented the maximum eigenvector coefficient in the 1ª component of the 2ª order watersheds: Ic only for the order analyzed (A) and Ff with fragmentation (B).

Source: author

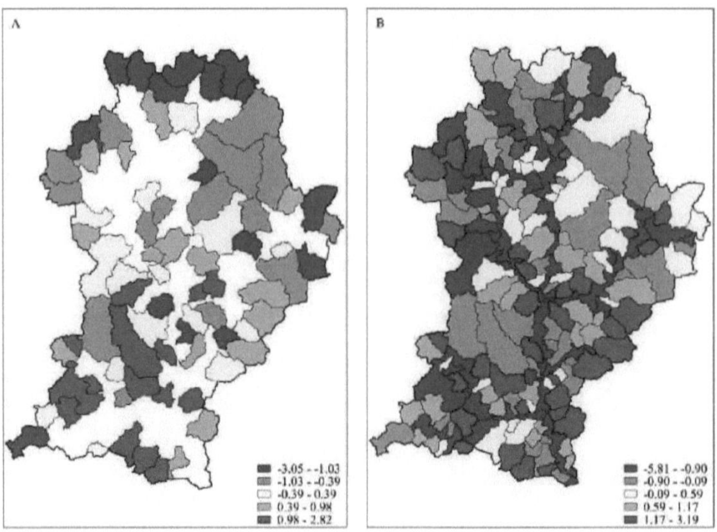

Figura 25 - ACP for the 2ª component of the 2ª order watersheds: only for the order analyzed (A) and with fragmentation (B).

Source: author

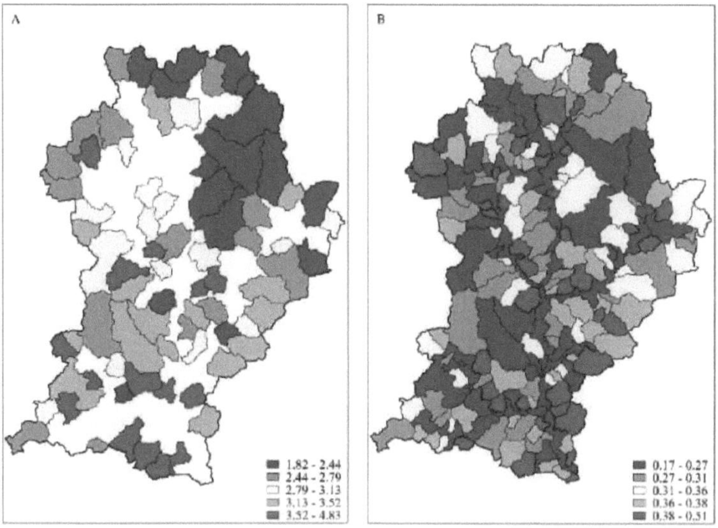

Figura 26 - Parameters that represented the maximum eigenvector coefficient in the 2ª component of the 2ª order watersheds: Gbm only for the order analyzed (A) and Ic with fragmentation (B).

Source: author

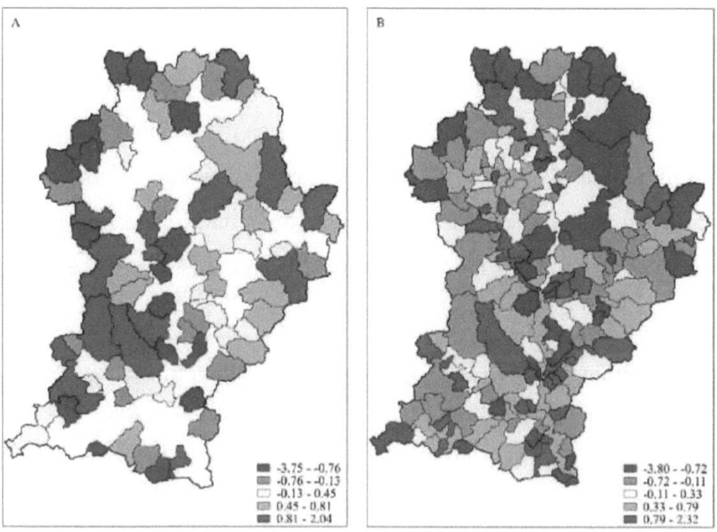

Figura 27 - ACP for the 3ª component of the 2ª order watersheds: only for the order analyzed (A) and with fragmentation (B).

Source: author

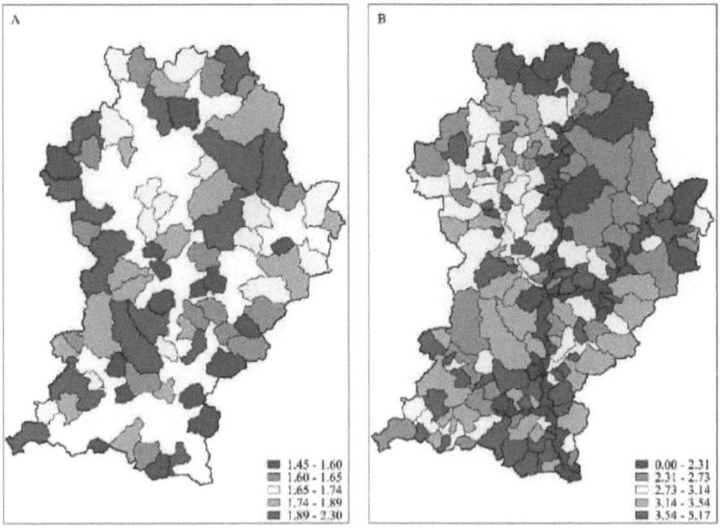

Figure 28 - Parameters that represented the maximum eigenvector coefficient in the 3ª component of the 2ª order watersheds: Kc only for the order analyzed (A) and Gbm with fragmentation (B).

Source: author

Analysis of the 3 watershedsª order

Tables 15 and 17 show all the correlations between the 3ª order watersheds, with and without a fragment, both between the parameters and between the parameters and the components, so that there was no nullity between the input parameters for the PCA, making it feasible to use it for analysis. The same tables show that Kc and Ic are fully correlated, as observed for the 1st (tables 6 and 9) and 2nd order (tables 11 and 13) watersheds.

Analyzing the results of the micro-watersheds without fragmentation, the parameters Ic, Gbm and Is were the parameters that had the greatest influence on the 1st, 2nd and 3rd components (table 16), a result that agrees with table 15, which shows that these same parameters had the greatest correlation with their respective components.

For the micro-basins with fragments, the parameters that most influenced the 1st, 2nd and 3rd components (Table 18) were Ff, Ic and Gbm, respectively.

these parameters with their respective components, as shown in Table 17.

Table 15 - Correlation matrix of morphometric parameters and principal components, only for the 3ª order watersheds analyzed. Only the first three principal components, which accumulated around 80% of the total variance of the data, were considered for analysis.

	Ff	Kc	Ic	Dd	Is	Gbm	dh	CP 1	CP 2	CP 3
Ff	**1.00**	**-0.68**	**0.68**	-0.13	0.13	-0.32	-0.41	**0.75**	0.18	-0.48
Kc	**-0.68**	**1.00**	-1.00	0.19	-0.33	-0.01	**0.50**	**-0.87**	-0.42	0.09
Ic	**0.68**	-1.00	**1.00**	-0.19	0.33	0.01	**-0.50**	0.87	0.42	-0.09
Dd	-0.13	0.19	-0.19	**1.00**	-0.31	**0.68**	**0.50**	**-0.52**	**0.76**	-0.06
Is	0.13	-0.33	0.33	-0.31	**1.00**	-0.02	**-0.51**	**0.53**	-0.09	0.74
Gbm	-0.32	-0.01	0.01	**0.68**	-0.02	**1.00**	0.26	-0.34	0.80	0.41
dh	-0.41	**0.50**	**-0.50**	**0.50**	**-0.51**	0.26	**1.00**	**-0.77**	0.21	-0.29
CP 1	**0.75**	**-0.87**	0.87	**-0.52**	**0.53**	-0.34	**-0.77**	**1.00**	0.00	0.00
CP 2	0.18	-0.42	0.42	**0.76**	-0.09	0.80	0.21	0.00	**1.00**	0.00
CP 3	-0.48	0.09	-0.09	-0.06		0.41	-0.29	0.00	0.00	**1.00**

Table 16 - Matrix of PCA eigenvector coefficients. Morphometric parameters that had the greatest weight in the variance of the data within each principal component, only for the 3ª order watersheds analyzed. Only the first three principal components, which accumulated around 80% of the total variance in the data, were considered for analysis.

	Ff	Kc	Ic	Dd	Is	Gbm	dh
CP 1	0.41	-0.48	0.48	-0.28	0.29	-0.18	-0.42
CP 2	0.14	-0.33	0.33	0.59	-0.07	0.62	0.17
CP 3	-0.47	0.08	-0.08	-0.06		0.40	-0.29

Table 17 - Correlation matrix of morphometric parameters and principal components for the 3ª watersheds plus fragments. Only the first three principal components, which accumulated around 80% of the total variance of the data, were considered for analysis.

	Ff	Kc	Ic	Dd	Is	Gbm	dh	CP 1	CP 2	CP 3
Ff	**1.00**	-0.42	0.42	**-0.67**	**-0.61**	-0.26	**-0.59**	0.87	-0.16	0.29
Kc	-0.42	**1.00**	-1.00	0.06	0.29	-0.01	0.32	**-0.69**	**-0.69**	0.17
Ic	0.42	-1.00	**1.00**	-0.06	-0.29	0.01	-0.32	**0.69**	0.69	-0.17
Dd	**-0.67**	0.06	-0.06	**1.00**	0.36	0.40	0.34	**-0.60**	**0.53**	-0.29
Is	**-0.61**	0.29	-0.29	**0.36**	**1.00**	0.09	0.33	**-0.65**	0.04	**-0.53**
Gbm	-0.26	-0.01	0.01	0.40	0.09	**1.00**	**0.59**	-0.40	**0.59**	0.55
dh	**-0.59**	0.32	-0.32	0.34	0.33	**0.59**	**1.00**	**-0.74**	0.28	0.40
CP 1	0.87	**-0.69**	**0.69**	**-0.60**	**-0.65**	**-0.40**	**-0.74**	**1.00**	0.00	0.00

CP 2	-0.16	**-0.69**	0.69	**0.53**	0.04	**0.59**	0.28	0.00	**1.00**	0.00
CP 3	0.29	0.17	-0.17	-0.29	**-0.53**		**0.40**	0.00	0.00	**1.00**

Table 18 - Matrix of PCA eigenvector coefficients. Morphometric parameters that had the greatest weight in the variance of the data within each component for the 3 watersheds[a] order plus fragments.

Only the first three principal components, which accounted for around 80% of the total variance in the data, were considered for analysis.

	Ff	**Kc**	**Ic**	**Dd**	**Is**	**Gbm**	**dh**
CP 1	0.49	-0.38	0.38	-0.33	-0.36	-0.24	-0.41
CP 2	-0.12	-0.53	0.53	0.40	0.03	0.46	0.21
CP 3	0.29	0.17	-0.17	-0.29	-0.54		0.40

The Ic values for the micro-watersheds without fragmentation (figure 30a) show that all the micro-watersheds have a high runoff potential, according to the classification in table 6, but it is still possible to identify the most circular micro-watersheds within the universe of analysis, i.e. the micro-watersheds with the highest Ic are more prone to flooding than the others. When compared with figure 29a, this pattern is partially lost.

As for the spatialization of the shape factor for the micro-watersheds with fragmentation (figure 30b), the evolution from circular to elongated patterns is more noticeable as the Ff decreases, precisely because of the presence of fragments. However, this evolution is clearer in the 1st principal component with fragmentation (figure 29b).

The spatial representation of the Gbm parameter (figure 32a), which represents more of the variance of the 2nd component for the micro-watersheds without a fragment, showed little similarity in spatial pattern when compared to its respective component (figure 31a).

The 2nd main component with fragmentation (figure 31b) proved to be unrepresentative in maintaining the spatial pattern of the parameter that most represented it, Ic (figure 32b), since the highest values of the component in question included more elongated watersheds, which is not consistent with higher Ic values.

The spatial arrangement of the 3rd component without fragmentation (figure 33a) did not correspond to the spatial pattern of the most representative parameter for the variance of the same component, Is (figure 34b).

The spatialization of the 3rd component with fragmentation (figure 33b) proved to be unrepresentative in maintaining the spatial pattern of the parameter that most represented it, Gbm (figure 34b).

As seen in the previous order analyses, the fragmented micro-basins made it possible to see the regionalization of both the components and the parameters with the highest eigenvector coefficients more clearly. [a]When comparing the three main components with their respective eigenvectors, which had the greatest influence on their variability, it is easier to see this relationship in the 1st component when spatializing them, with the greatest influence of the Ff in the 3rd order, followed by the Ic for the micro-basins with fragments (Table 18).

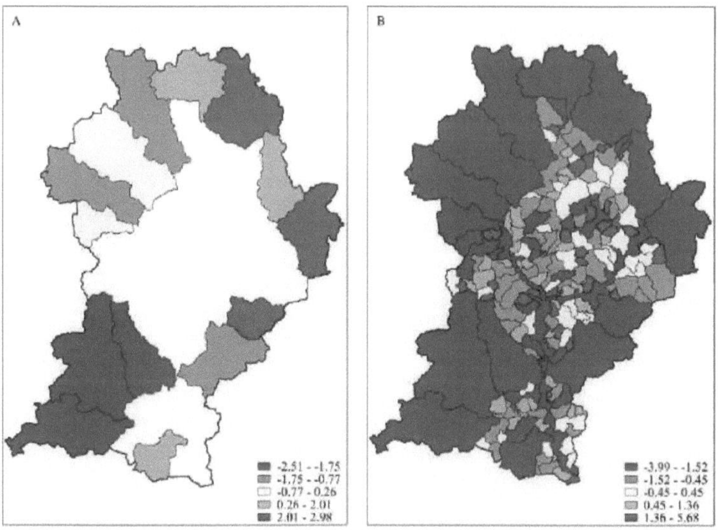

Figure 29 - PCA for the P component of the 3rd order watersheds: only for the order analyzed (A) and with fragmentation (B).

Source: author

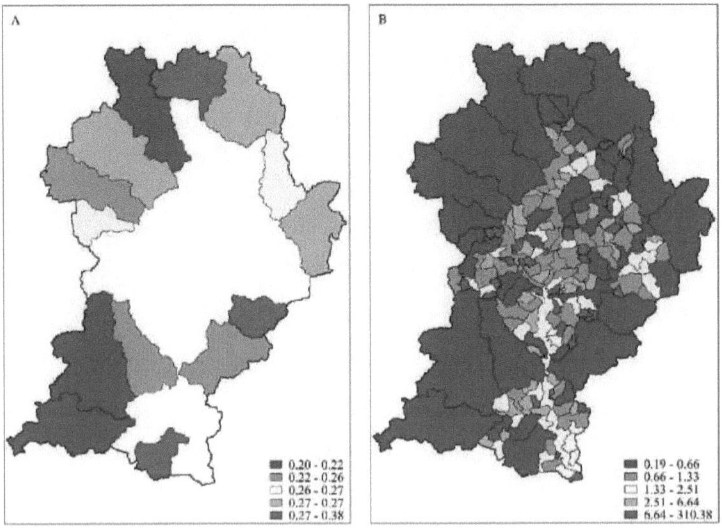

Figura 30 - Parameters that represented the maximum eigenvector coefficient in the 1ª component of the 3ª order watersheds: Ic only for the order analyzed (A) and Ff with fragmentation (B).

Source: author

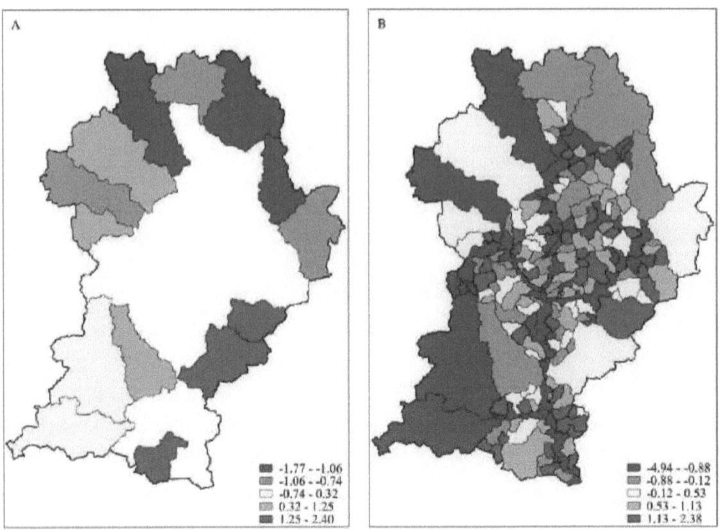

Figura 31 - ACP for the 2ª component of the 3ª order watersheds: only for the order analyzed (A) and with fragmentation (B).

Source: author

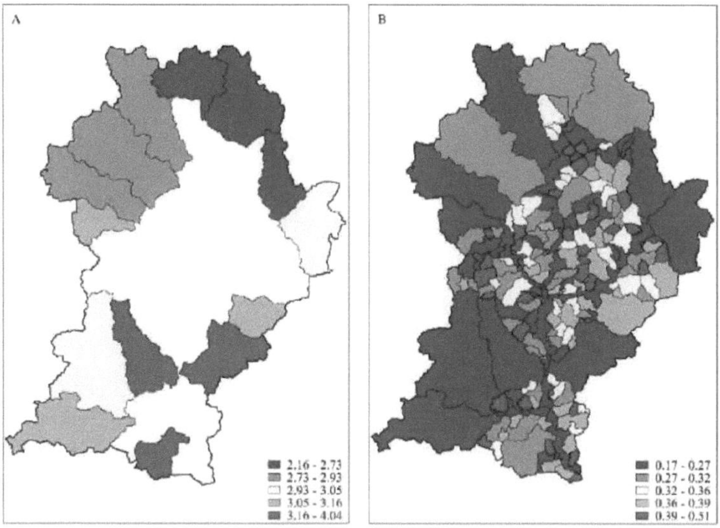

Figura 32 - Parameters that represented the maximum eigenvector coefficient in the 2ª component of the 3ª order watersheds: Gbm only for the order analyzed (A) and Ic with fragmentation (B).

Source: author

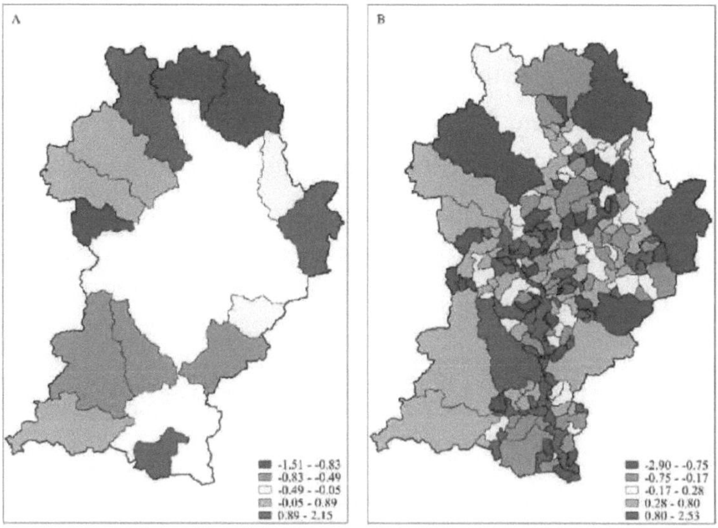

Figura 33 - ACP for the 3ª component of the 3ª order watersheds: only for the order analyzed (A) and with fragmentation (B).

Source: author

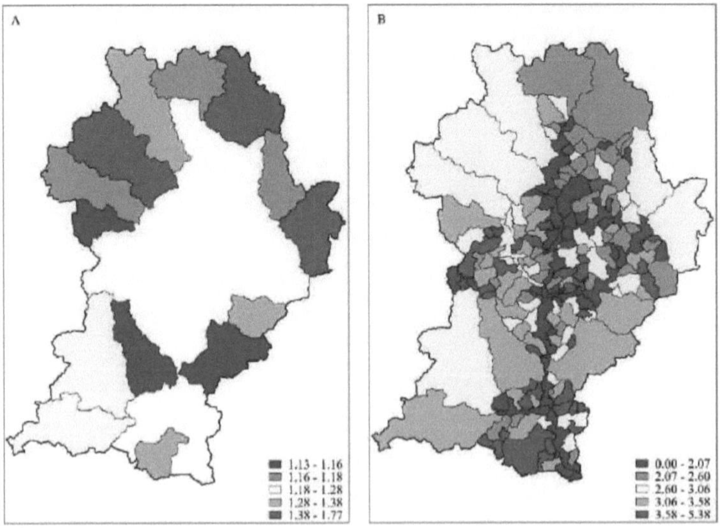

Figure 34 - Parameters that represented the maximum eigenvector coefficient in the 3ª component of the 3ª order watersheds: Is only for the order analyzed (A) and Gbm with fragmentation (B).

Source: author

Analysis of the 4 watershedsª order

Tables 19 and 21 show all the correlations between the 4ª order watersheds, with and without fragments, both between the parameters and between these and the components, so that there was no nullity between the input parameters for the PCA, making it feasible to use it for analysis. The same tables show that Kc and Ic are fully correlated, as observed for the 1st order (tables 7 and 9), 2nd order (tables 11 and 13) and 3rd order (tables 15 and 17) watersheds.

In Table 19, the parameters Kc and Ic are again fully correlated, and they also show a high correlation with dh for the micro-watersheds without fragmentation, so that Kc and dh are directly proportional quantities (as one increases, the other also increases) and Ic and dh are inversely proportional (as one increases, the other decreases).

Table 19 also shows that the correlation values, for example for Ic and dh, are not in line with the pattern observed in previous analyses. This may be due to the number of watersheds analyzed, which was only 4, Therefore, in order to consider the parameter that most influenced each component, Table 20 was taken into account, with the parameters dh, Gbm and Is contributing the most to the variance of the 1st, 2nd and 3rd components.

It is worth remembering that the same thing happened for the 2nd[a] order micro-basins without fragments (table 11), and also here for the 4th order micro-basins without fragments. It can be inferred that the lack of information in the analysis only for the micro-basins without fragments could jeopardize the results of the study of morphometric parameters.

The parameters Ff, Gbm and dh (table 22) had the greatest influence on the 1st, 2nd and 3rd components, respectively, which is in line with the highest correlations between the parameters and their respective components (table 21).

Table 19 - Correlation matrix of morphometric parameters and principal components, only for the 4th order watersheds analyzed. Only the first three principal components, which accumulated around 80% of the total variance of the data, were considered for analysis.

	Ff	Kc	Ic	Dd	Is	Gbm	dh	CP 1	CP 2	CP 3
Ff	**1.00**	0.07	-0.07	0.45	0.22	-0.43	-0.05	0.04	-1.00	-0.08
Kc	0.07	**1.00**	-1.00	0.49	**0.77**	**0.76**	0.92	0.98	-0.04	0.18
Ic	-0.07	-1.00	**1.00**	-0.49	**-0.77**	**-0.76**	-0.92	-0.98	0.04	-0.18
Dd	0.45	0.49	-0.49	**1.00**	-0.02	**0.51**	**0.69**	0.62	-0.37	**-0.70**
Is	0.22	**0.77**	**-0.77**	-0.02	**1.00**	0.24	0.46	0.64	-0.25	**0.73**
Gbm	-0.43	**0.76**	**-0.76**	**0.51**	0.24	**1.00**	0.92	**0.84**	0.48	-0.25
dh	-0.05	0.92	-0.92	**0.69**	0.46	0.92	**1.00**	0.98	0.10	-0.19
CP 1	0.04	0.98	-0.98	**0.62**	**0.64**	**0.84**	0.98	**1.00**	0.00	0.00
CP 2	-1.00	-0.04	0.04	-0.37	-0.25	0.48	0.10	0.00	**1.00**	0.00
CP 3	-0.08	0.18	-0.18	**-0.70**		-0.25	-0.19	0.00	0.00	**1.00**

Table 20 - Matrix of PCA eigenvector coefficients. Morphometric parameters that had the greatest weight in the variance of the data within each principal component, only for the 4th order watersheds analyzed. Only the first three principal components, which accumulated around 80% of the total variance in the data, were considered for analysis.

	Ff	Kc	Ic	Dd	Is	Gbm	dh
CP 1	0.02	0.47	-0.47	0.29	0.31	0.40	0.47
CP 2	-0.83	-0.03	0.03	-0.31	-0.21	0.40	0.09

CP 3	-0.07	0.17	-0.17	-0.64		-0.23	-0.17

Table 21 - Correlation matrix of morphometric parameters and principal components for the 4ª order micro basins plus fragments. Only the first three principal components, which accumulated around 80% of the total variance of the data, were considered for analysis.

	Ff	**Kc**	**Ic**	**Dd**	**Is**	**Gbm**	**dh**	**CP 1**	**CP 2**	**CP 3**
Ff	**1.00**	-0.46	0.46	**-0.75**	**-0.55**	-0.25	-0.44	0.86	-0.21	0.30
Kc	-0.46	**1.00**	-1.00	0.11	0.26	-0.06	0.39	**-0.75**	**-0.63**	0.13
Ic	0.46	-1.00	**1.00**	-0.11	-0.26	0.06	-0.39	**0.75**	**0.63**	-0.13
Dd	**-0.75**	0.11	-0.11	**1.00**	0.37	0.36	0.22	**-0.60**	**0.54**	-0.37
Is	**-0.55**	0.26	-0.26	0.37	**1.00**	0.06	0.23	**-0.59**	0.11	-0.49
Gbm	-0.25	-0.06	0.06	0.36	0.06	**1.00**	**0.56**	-0.32	0.66	**0.53**
dh	-0.44	0.39	-0.39	0.22	0.23	**0.56**	**1.00**	**-0.67**	0.19	0.58
CP 1	0.86	**-0.75**	**0.75**	**-0.60**	**-0.59**	-0.32	**-0.67**	**1.00**	0.00	0.00
CP 2	-0.21	**-0.63**	**0.63**	**0.54**	0.11	0.66	0.19	0.00	**1.00**	0.00
CP 3	0.30	0.13	-0.13	-0.37	-0.49	**0.53**		0.00	0.00	**1.00**

Table 22 - Matrix of PCA eigenvector coefficients. Morphometric parameters that had the greatest weight in the variance of the data within each component for the 4ª order micro-basins plus fragments. Only the first three principal components, which accumulated around 80% of the total variance in the data, were considered for analysis.

	Ff	**Kc**	**Ic**	**Dd**	**Is**	**Gbm**	**Dh**
CP 1	0.48	-0.42	0.42	-0.34	-0.33	-0.20	-0.38
CP 2	-0.17	-0.49	0.49	0.42	0.09	0.53	0.15
CP 3	0.28	0.12	-0.12	-0.35	-0.46	0.51	

The dh values for the micro-watersheds without a fragment (figure 36a) are well represented by their respective component for the micro-watersheds without a fragment (figure 35a), which may be due to the fact that there are few micro-watersheds represented.

The spatial arrangement of the 1st component for micro-watersheds with fragmentation (figure 35b) was similar to that presented by the Ff parameter for micro-watersheds with fragments, which represented the component's greatest variability (figure 36b).

The spatialization of the 2nd component for micro-watersheds without fragments (figure 37a) did not have a similar spatial arrangement to that presented by the Gbm parameter for micro-watersheds without fragments, which represented the component's greatest variability (figure 38a).

The distribution pattern of the micro-watersheds with fragmentation of the Gbm parameter (figure 38b) along the flatter part of the terrain is clear, and this behavior can also be seen in the spatialization of the 2nd component for micro-watersheds with fragmentation (figure 37b).

The spatialization of the 3rd component for micro-watersheds without fragmentation (figure 39a) partially preserved the spatial characteristic of the Is parameter (figure 40a), which most influenced the variance of the component in question.

For the spatialization of the 3rd component for the micro-watersheds with fragmentation (figure 39b), the spatial arrangement of the dh parameter (figure 40b), which most influenced the variance of the component, was not preserved.

As seen in all the previous analyses, the fragmented watersheds made it possible to see the regionalization of both the components and the parameters with the highest eigenvector coefficients more clearly. When comparing the three main components with their respective eigenvectors, which had the greatest influence on their variability, it is easier to see this relationship in the 1st component when spatializing them, with the Ff having the greatest influence for the 4th[a] order, followed by the Ic for the micro-basins with fragments (Table 22).

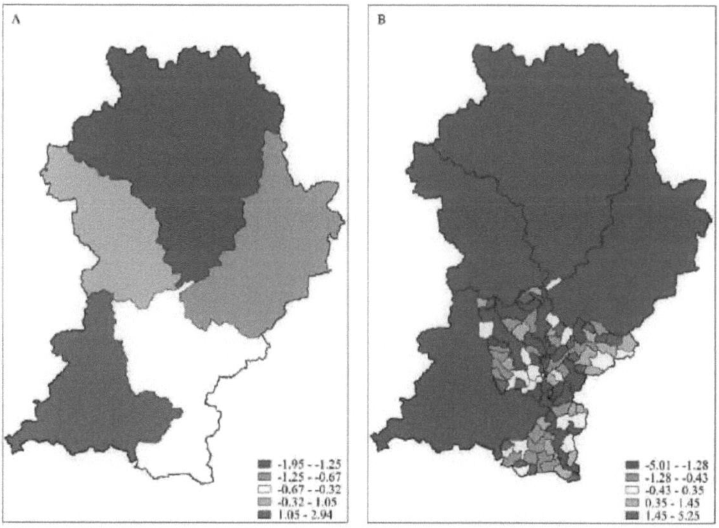

Figure 35 - PCA for the P component of the 4th order watersheds: only for the order analyzed (A) and with fragmentation (B).

Source: author

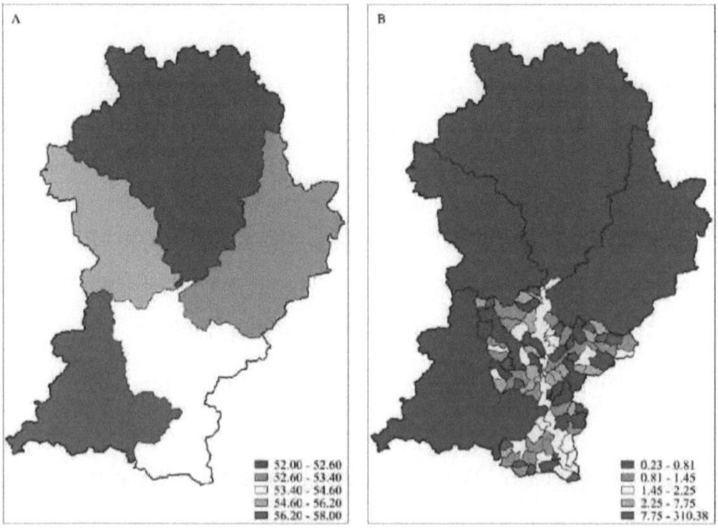

Figura 36 - Parameters that represented the maximum eigenvector coefficient in the 1ª component of the 4ª order watersheds: dh only for the order analyzed (A) and Ff with fragmentation (B).

Source: author

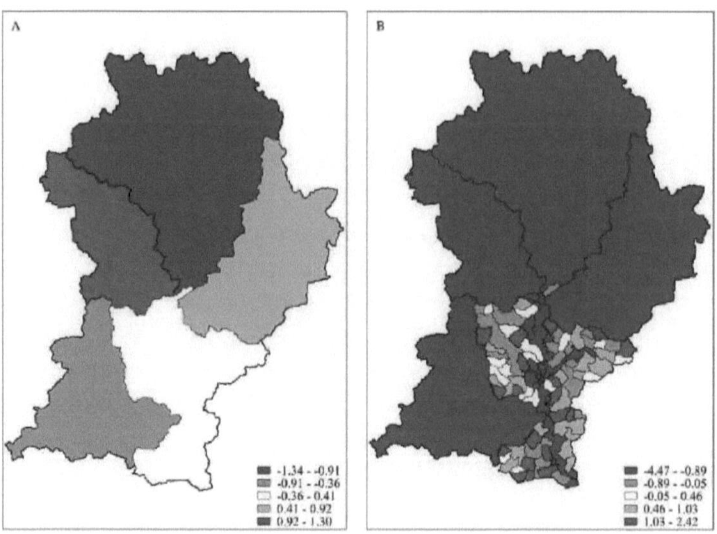

Figura 37 - ACP for the 2ª component of the *4ª* order watersheds: only for the order analyzed (A) and with

fragmentation (B).

Source: author

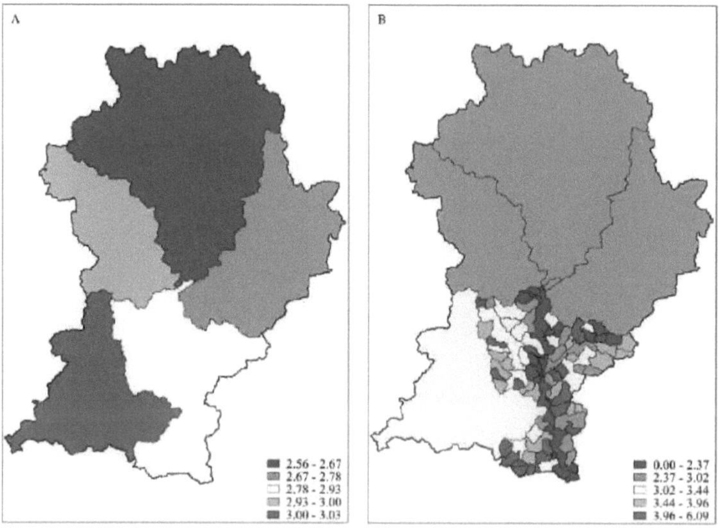

Figura 38 - Parameters representing the maximum eigenvector coefficient in the 2ª component of the 4ª order watersheds: Gbm only for the order analyzed (A) and Gbm with fragmentation (B).

Source: author

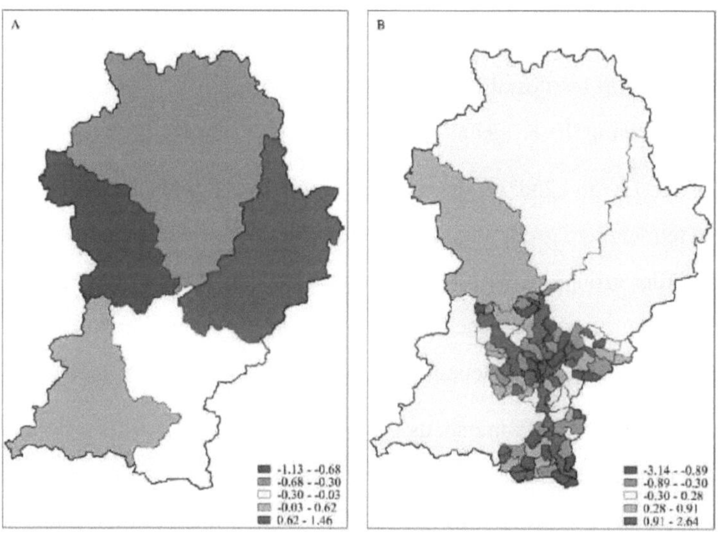

Figura 39 - PCA for the 3ª component of the *4ª* order watersheds: only for the order analyzed (A) and with fragmentation (B).

Source: author

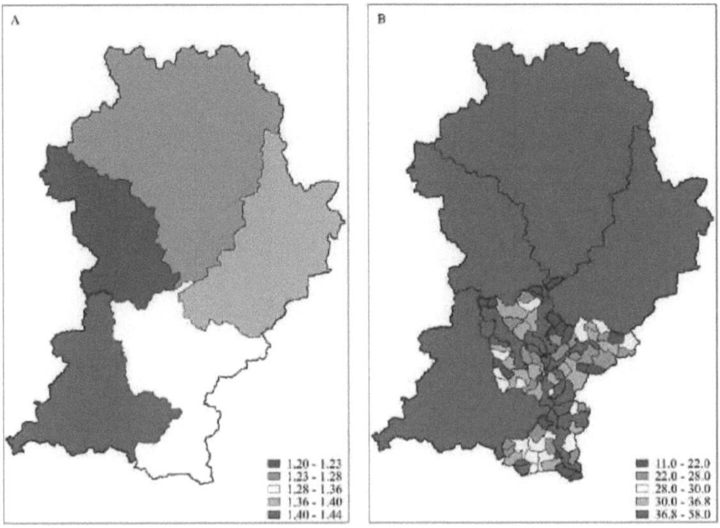

Figure 40 - Parameters that represented the maximum eigenvector coefficient in the 3ª component of the 4ª order watersheds: Is only for the order analyzed (A) and dh with fragmentation (B).

Source: author

Grouping of watersheds with similar morphometric characteristics.

In order to identify similar groups of watersheds with the aim of delimiting management units to support territorial management of the Apeù river basin, a cluster analysis was carried out using the K-means method, and the results are shown below:

For the groupings of the 1stª and 2ndª orders with fragmentation (figures 41b and 42b), we can see a certain tendency to group the lower areas with flatter relief (figures 6 and 7). The 1st and 2nd order groupings without fragmentation, shown in figures 41a and 42a, show that the shape criterion of these micro-basins was the most relevant, where the more elongated micro-basins were separated from the more circular micro-basins.

For the groupings of the 3rd order watersheds (with and without fragments - figures 43a and 43b) and 4th order watersheds with fragments (figure 44b), it was not possible to visually identify a spatial pattern. It was not possible to generate a cluster analysis for the 4th order watersheds without fragments (figure 44a), as the number of watersheds for this method was insufficient to carry out the procedure.

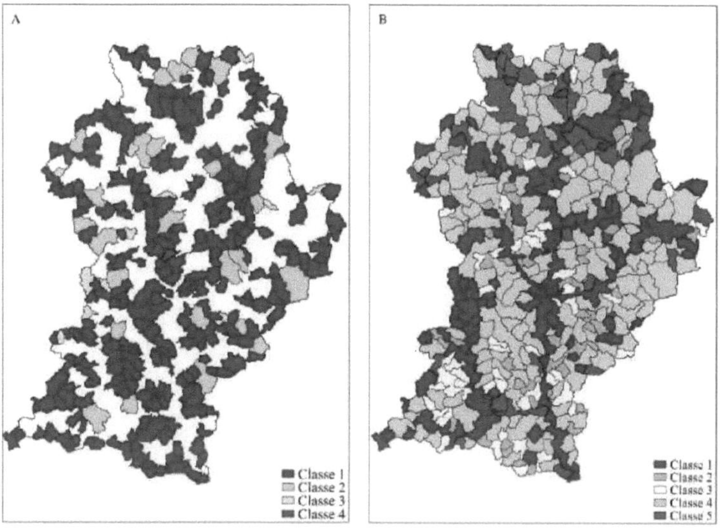

Figura 41 - Classification using the K-means method for 1ª order watersheds only (A) and for 1ª order watersheds plus fragments (B).

Source: author

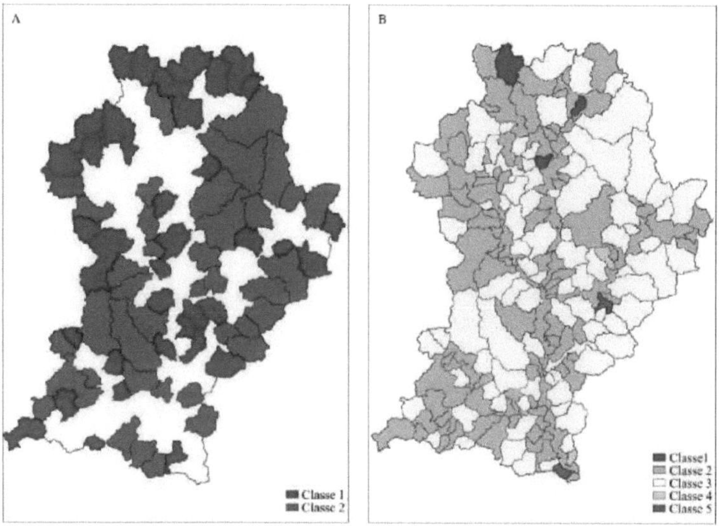

Figura 42 - Classification by the K-means method for the 2nd order micro-watersheds only[a] (A) and for the 2nd order micro-watersheds plus fragments (B).

Source: author

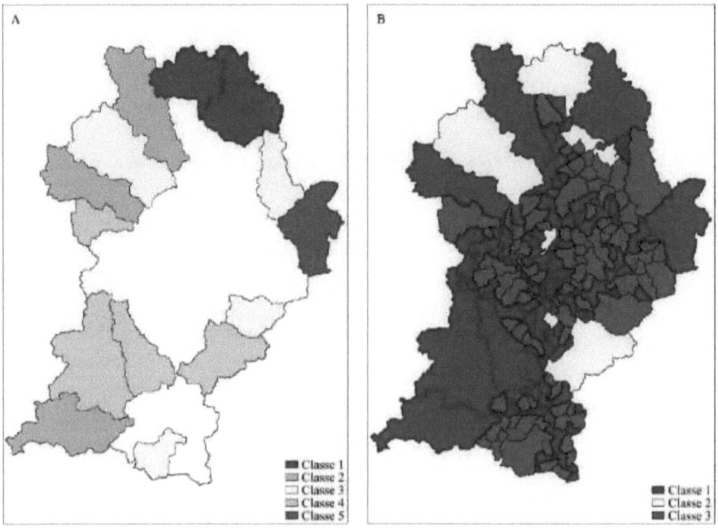

Figura 43 - Classification using the K-means method for 3ª order watersheds only (A) and for 3ª order watersheds plus fragments (B).

Source: author

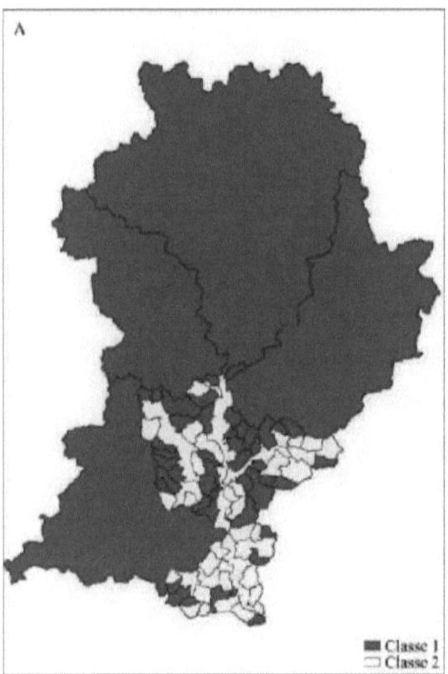

Figura 44 - (A) and classification by the K-means method for the *4ª* order micro-watersheds plus fragments.

Source: author

As each sub-treatment had a parameter that most influenced each principal component, it was then decided to check which parameter most influenced the total variability of the data from the sum of their absolute frequency. The result is shown in Table 23.

Table 23 - Parameters that concentrated the greatest variability in the data.

Ranking	Parameters	∑ Coef. Eigenvector
1°	Gbm	7
2°	Ff	5
3°	Ic	4
4°	Kc	3
5°	dh	3
6°	Is	2
7°	Dd	0

The parameters that most influenced the total variability of the data were Gbm, Ff and Ic. Gbm is a parameter that characterizes the terrain under study, revealing the type of relief in which the basin is located, which is important information for morphometric studies. Ff refers to the shape of the basins, classifying them into different flood risks. The Ic reveals the potential for surface runoff, as well as the rate of water infiltration into the soil (CHRISTOFOLLETI, 1980). Therefore, the ranking of the parameters revealed characteristics that complement each other and can give a good idea of the characteristics of a particular area, especially when there is little information about it.

CONCLUSION

The methodology for obtaining the hydrologically corrected Digital Elevation Model allowed for an accurate automatic delimitation of the drainage micro-basins and the hydrographic network of the Apeù river basin.

The Apeù river basin is classified as 5ª order, with relief varying from flat to strongly undulating and an average slope of 3%, as well as having a more elongated shape and high runoff potential, characterizing a basin with a low risk of flooding.

The application of Horton's laws to the modeled drainage and watershed data made it possible to verify the validity of the watershed modeling method by means of the TauDEM algorithm using the MDE corrected for the drainage network as an *input*.

Obtaining the morphometric parameters made it possible to apply the analysis of

Principal Components, showing the relevance of the 1st principal component in representing the shape factor (Ff) and circularity index (Ic) parameters, respectively, in the variability of the morphometric data of the basin in question.

The spatial patterns were better evidenced when the micro-basins of the different orders were analyzed with the fragments.

The parameters Average Basin Slope, Shape Factor and Circularity indices had the greatest influence on the total variability of the morphometric data.

The clustering analysis made it possible to generate management units with similar morphometric characteristics, as well as to define the relevant parameters in the formation of these units, thus subsidizing future territorial planning and management actions in the Apeù river basin.

REFERENCES

ARAÙJO, J. P. DE C.; CASTRO, T. G. S. DE; SILVA, A. C. DA; MACHADO, P. J. DE O.; ZAIDAN, R. T. Application of methods and criteria for determining a main river: the Paraibuna River basin. **XIII Brazilian Symposium on Geography**, p. 18, 2007. Available at:

<http://www.geomorfologia.ufv.br/simposio/simposio/trabalhos/trabalhos_completos/eixo1/040.pdf >. Accessed on: 03 Jan. 2017.

BERNARDI, E. C. S.; PANZIERA, A. G.; BURIOL, G. A.; SWAROWSKY, A. Hydrographic Basin as an Environmental Management Unit. **Natural and Technological Sciences**, v. 13, p. 159168, 2013. Available at: <http://sites.unifra.br/Portals/36/Tecnologicas 2012-2/04.pdf>. Accessed on: Feb. 22, 2017.

BRAGA, B.; HESPANHOL, I.; CONEJO, J. G. L.; et al. **Introduction to Environmental Engineering**. 2ª ed. Sao Paulo - SP: Pearson Prentice Hall, 2005.

CARDOSO, C. A.; DIAS, H. C. T.; SOARES, C. P. B.; MARTINS, S. V. Morphometric characterization of the Debossan river basin, Nova Friburgo, RJ. **Revista Arvore**, v. 30, n. 2, p. 241-248, 2006. Available at:

<http://www.scielo.br/scielo.php?script=sci_arttext&pid=S0100-

67622006000200011&lng=pt&nrm=iso&tlng=pt>. Accessed on: 04 Jan. 2017.

CARMO, J. P. DE A. DO; SILVA, P. D. D. The Hydrographic Basin as a Unit for Study, Planning and Management. **XVI Encontro Nacional dos Geógrafos**, v. 2, p. 1-9, 2010. Available at: <http://www.agb.org.br/evento/download.php?idTrabalho=4055>. Accessed on: Feb. 20, 2017.

CHEREM, L. F. S. **Anâlise Morfométrica da Bacia do Alto Rio das Velhas - MG Anâlise**, 2008. Federal University of Minas Gerais. Available at:

<http://csr.ufmg.br/modelagem/dissertacoes/luizfelipecherem.pdf>. Accessed on: 05 Jan. 2017.

CHORLEY, R. J. Geomorphology and general systems theory. **US Geological survey professional paper 500-B**, p. B1-B10, 1962. Available at: <https://pubs.usgs.gov/pp/0500b/report.pdf>.

Accessed on: January 19, 2017.

CHORLEY, R. J.; KENNEDY, B. A. Physical geography: a systems approach. **Quarterly Journal of the Royal Meteorological Society**, v. 98, n. 418, p. 370, 1972. Englewood Cliffs - NJ: PrenticeHall International. Disponivel em: <http://doi.wiley.com/10.1002/qj.49709841818>. Accessed on: 02 Dec. 2016.

CHRISTOFOLETTI, A. **Geomorfologia**. 2ª ed. Sao Paulo - SP: Editora Edgard Blucher Ltda, 1980.

COLLARES, E. G. **Evaluation of Changes in Micro-basin Drainage Networks as a Subsidiary to the Geo-environmental Zoning of Hydrographic Basins: Application to the Capivari River Basin - SP**, 2000. University of Sao Paulo. Available at: <http://www.teses.usp.br/teses/disponiveis/18/18132/tde-25012001-115727/publico/Collares.pdf>. Accessed on: Dec. 14, 2016.

EMBRAPA. **Slope classes**. Rio de Janeiro, 1979.

FLORENZANO, T. G. **Geomorfologia: conceitos e tecnologias actuais**. 2ª ed. Sao Paulo - SP: Oficina de Textos, 2008.

FONSECA, B. M. **O Uso do Sistema de Informaçôes Geogràficas da Anâlise Morfométrica d Morfológica de Bacias de Drenagem da Serra do Espinhaço Meridional - Mg**, 2010. Federal University of Minas Gerais. Available at:

<http://www.bibliotecadigital.ufmg.br/dspace/handle/1843/MPBB-86XLMN>. Accessed on: 15 Feb. 2017.

GÜNAL, A. Y.; GÜVEN, A. Determination of Geomorphological Parameters of Damlica Basin Using GIS. **Acta Physica Polonica A**, v. 128, n. 2, p. 222-224, 2015. Available at: <http://przyrbwn.icm.edu.pl/APP/PDF/128/a128z2bp063.pdf>. Accessed on: Jan. 19, 2017.

HORTON, R. E. Erosional Development of Streams and their Drainage Basins; Hydrophysical Approach to Quantitative Morphology. **Geological Society of America Bulletin**, v. 56, n. 1, p. 151-180, 1945. Available at: <http://www.geos.ed.ac.uk/homes/s0451705/horton_1945.pdf>.

Accessed on: 20 Jan. 2017.

HOSSEINZADEH, S. R. Drainage Network Analysis , Comparis of Digital Elevation Model (DEM) from ASTER with High Resolution Satellite Image and Areal Photographs. **International Journal of Environmental Science and Development**, v. 2, n. 3, p. 194-198, 2011. Available 75

at: <http://www.ijesd.org/papers/123-S030.pdf>. Accessed on: 08 Jan. 2017.

LI, Z. Watershed modeling using arc hydro based on DEMs : a case study in Jackpine watershed. **Li Environmental Systems Research**, p. 1-12, 2014. Available at:

<https://link.springer.com/article/10.1186/2193-2697-3-11>. Accessed on: 12 Feb. 2017.

OLIVEIRA, S. N. DE. **Methodology for delimitation and multivariate analysis of morphometric attributes of a drainage basin using a hydrologically corrected digital elevation model**, 2008. UNIVERSIDADE DE BRASILIA. Available at:

<http://repositorio.unb.br/handle/10482/3511>. Accessed on: 04 Feb. 2017.

ROCHA, A. A.; VIANNA, P. C. G. The Hydrographic Basin as a Water Management Unit. **II SEMILUSO - Seminàrio Luso-Brasileiro Agricultura Familiar e Desertificaçâo**, p. 1-11, 2008. Available at: <http://www.geociencias.ufpb.br/leppan/gepat/files/gepat022.pdf>. Accessed on: 05 Jan. 2017.

ROCHA, K. J. DA. **Composition and structure of floristic groups in a secondary forest fragment**, 2015. Federal University of Mato Grosso. Available at:

<http://www.ufmt.br/ufmt/unidade/userfiles/publicacoes/2cfc5af9dd27e270fb65761c50212ad7.pdf >. Accessed on: 25 Jan. 2017.

SALGADO, M. P. G.; BATISTA, G. T.; DIAS, N. W.; TARGA, M. DOS S. Characterization of a watershed through geotechnologies. **XIV Brazilian Symposium on Remote Sensing**, p. 4837-4843, 2009. Available at:

<http://marte.dpi.inpe.br/rep/dpi.inpe.br/sbsr@80/2008/11.17.19.32?languagebutton=pt-BR&mirror=dpi.inpe.br/banon/2003/12.10.19.30.54&searchsite=bibdigital.sid.inpe.br:80&searchmirror=sid.inpe.br/bibdigital@80/2006/04.07.15.50.13>. Accessed on: 22 Jan. 2017.

SANTOS, A. M. DOS; TARGA, M. DOS S.; BATISTA, G. T.; DIAS, N. W. Morphometric analysis of the Perdizes and Fojo sub-basins in the municipality of Campos do Jordao, SP, Brazil. **Revista Ambiente e Agua**, v. 7, n. 1980-993X, p. 195-211, 2012. Available at:

<http://www.scielo.br/pdf/ambiagua/v7n3/v7n3a16.pdf>. Accessed on: 16 Feb. 2017.

SANTOS, D. A. R.; MORAIS, F. Morphometric analysis of the Lago Verde River Basin as a subsidy to the compartmentalization of the relief of the region of Lagoa da Confusao - TO. **Revista Geonorte, Special Issue**, v. 3, n. 4, p. 617-629, 2012. Available at: <http://www.periodicos.ufam.edu.br/revista-geonorte/article/viewFile/1975/1849>. Accessed on: 09 Jan. 2017.

SANTOS, O. C. DE O. Anâlise do Uso do Solo e dos Recursos Hidricos na Microbacia do Igarapé Apeù - Nordeste do Estado do Parâ. **Anuârio do Intituto de Geociências - UFRJ**, v. 29, n. 2, p. 225-246, 2006. Available at:

<http://igc.usp.br/igcJournal/index.php/anigeo/article/view/4883/4395>. Accessed on: 13 Feb. 2017.

SCHMITT, A.; MOREIRA, C. R. Watershed management using the free Quantun-GIS software. **Revista Cultivando o Saber**, p. 125-137, 2015. Available at:

<http://www.fag.edu.br/upload/revista/cultivando_o_saber/566ec59f31e47.pdf>. Accessed on: 04 Feb. 2017.

SCHUMM, S. A. Evolution of Drainage Systems and Slopes in Badlands at Perth Amboy. **Geological Society of America Bulletin**, v. 67, n. 5, p. 597-646, 1956. Available at:

<http://gsabulletin.gsapubs.org/cgi/doi/10.1130/0016-7606(1956)67[597:EODSAS]2.0.CO;2>. 76

SILVA, A. A.; SANTOS, G. O.; BRAZ, A. R. C. **Morphometric characterization of watersheds in the municipality of rio verde, goiàs**. Rio Verde - GO, 2015.

SOARES, M. R. G. D. J.; SOUZA, J. L. M. DE. Morphometric analysis of the rio pequeno hydrographic basin in sao josé dos pinhais (PR). **Revista Geografia (Londrina)**, v. 20, n. 3, p. 57-70, 2012. Available at: <http://www.uel.br/revistas/uel/index.php/geografia/article/view/8982/12668>. Accessed on: 05 Feb. 2017.

SODRÉ, R. V. R.; JÙNIOR, O. A. DE C.; MACHADO, W. P.; et al. Classification of drainage basins in Alto Jequitai (Minas Gerais) using principal component analysis and group analysis. **Revista Brasileira de Geomorfologia**, v. 2, p. 73-86, 2007. Available at:

<http://www.lsie.unb.br/rbg/index.php?journal=rbg&page=article&op=view&path%5B%5D=95>. Accessed on: 17 Jan. 2017.

SOUZA, S. R. DE; MACIEL, M. DE N. M.; OLIVEIRA, F. DE A.; JESUINO, S. DE A. Characterization of land use and occupation conflicts in the permanent preservation areas of the Apeù River, northeastern Parà. **Revista Florestas**, , n. 2005, p. 701-710, dec. 2012. Curitiba. Available at:

<https://www.google.com.br/url?sa=t&rct=j&q=&esrc=s&source=web&cd=1&ved=0ahUKEwi0yKvXo9vSAhXCHZAKHYdOBL0QFggcMAA&url=http%3A%2F%2Frevistas.ufpr.br%2Ffloresta%2Farticle%2Fdownload%2F19508%2F19828&usg=AFQjCNHIqOgdJqChzeldpA1eBoawUSNnZQ&sig2=_3YpeaRan0fcgkVwHw9PLg>. Accessed on: January 18, 2017.

STRAHLER, A. N. Hypsometric (Area-Altitude) Analysis of Erosional Topography. **Bulletin of the Geological Society of America**, v. 63, p. 1117-1142, 1952. Available at: <http://www.unc.edu/courses/2010spring/geog/591/001/students/nmey13/GEOL483/Lab5/pdfs/Strahler_1952_hypsometry.pdf>. Accessed on: Feb. 24, 2017.

STRAHLER, A. N. Quantitative Analysis of Watershed Geomorphology. **American Geophysical Union**, v. 38, p. 913-920, 1957. Available at:

<http://www.uvm.edu/pdodds/files/papers/others/1957/strahler1957a.pdf>. Accessed on: 22 Feb. 2017.

STRAHLER, A. N. Dimensional Analysis Applied to Fluvially Eroded Landforms. **Geological Society of America Bulletin**, v. 69, n. 3, p. 279-300, 1958. Available at:

<http://gsabulletin.gsapubs.org/cgi/doi/10.1130/0016-7606(1958)69[279:DAATFE]2.0.CO;2>.

Accessed on: 02 Feb. 2017.

TARBORON, D. G. A new method for the determination of flow directions and upslope areas in grid digital elevation models. **Water Resources Research**, v. 33, n. 2, p. 309-319, 1997.

Available at: <http://onlinelibrary.wiley.com/doi/10.1029/96WR03137/pdf>. Accessed on: 03 Jan. 2017.

TARBOTON, D. G.; MOHAMMED, I. N. Quick Start Guide to Using the Taudem Arcgis Toolbox. Available at: <http://hydrology.usu.edu/taudem/taudem5/TauDEM51GettingStartedGuide.pdf>. Accessed on: 10/12/2016.

TEODORO, V. L. I.; TEIXEIRA, D.; COSTA, D. J. L.; FULLER, B. B. The Concept of Hydrographic Basin and the Importance of Morphometric Characterization for Understanding Local Environmental Dynamics. **Revista Uniara**, v. 20, p. 136-156, 2007. Available at:

<http://www.uniara.com.br/legado/revistauniara/pdf/20/RevUniara20_11.pdf>. Accessed on: January 11, 2017.

TONELLO, K. C. **Anâlise Hidroambiental da Bacia Hidrogràfica da Cachoeira das Pombas, Guanhaes, Mg**, 2005. Federal University of Viçosa. Available at:

<http://vampira.ourinhos.unesp.br:8080/cediap/material/analise_hidroambiental_em_microbacias.pdf>. Accessed on: January 24, 2017.

TUCCI, C. E. M. **Hidrologia: ciência e aplicçao**. 2° ed. Porto Alegre - RS: ABRH, 1993.

VALERIANO, M. D. M.; MORAES, J. F. L. DE. Extraction of drainage networks and dividers by digital processing of topographic data. **X Brazilian Symposium on Remote Sensing**, p. 517-524, 2001. Available at:

<http://marte.sid.inpe.br/col/dpi.inpe.br/lise/2001/09.19.07.26/doc/0517.524.084.pdf>. Accessed on: 05 Jan. 2017.

VALERIANO, M. DE M. **Digital elevation models of micro-basins developed with kriging**. Sao José dos Campos, 2002.

VALERIANO, M. DE M. Digital model of morphometric variables with SRTM data for the national territory: the TOPODATA project. **XII Brazilian Symposium on Remote Sensing**, p. 3595-3602, 2005. Available at:

<http://marte.sid.inpe.br/col/ltid.inpe.br/sbsr/2004/10.29.11.41/doc/3595.pdf>. Accessed on: 08 Feb. 2017.

VALERIANO, M. DE M.; ALBUQUERQUE, P. C. G. DE. Derivation of Digital Elevation Models

(DEM) in geographic coordinates. **XVII Brazilian Symposium on Remote Sensing**, p. 1860-1867, 2015. Available at: <http://www.dsr.inpe.br/sbsr2015/files/p0369.pdf>. Accessed on: 04 Jan. 2017.

VALERIANO, M. DE M.; ROSSETTI, D. DE F. **TOPODATA: selection of geostatistical coefficients for the unified refinement of SRTM data**. Sao José dos Campos, 2010.

VEIGA, A. M.; TRINDADE, M. C.; SOUZA, R. M. DE; OLIVEIRA, W. N. DE. Hydromorphological Characterization of the Rio dos Bois Basin. **XX Simpòsio Brasileiro de Recursos Hidricos**, , n. 62, p. 1-7, 2013. Available at:

<http://www.cprm.gov.br/publique/media/Evento_Veiga_Carac.pdf>. Accessed on: Feb. 19, 2017.

VESTENA, L. R.; CHECCHIA, T.; KOBIYAMA, M. Morphometric analysis of the Caeté watershed, Alfredo Wagner/SC. **VI National Symposium on Geomorphology**, p. 1-8, 2006. Available at: <http://www.labhidro.ufsc.br/Artigos/Vestena et al. - Morphometric analysis ... - VI SINAGEO-1.pdf>. Accessed on: 03 Jan. 2017 Accessed on: 03 Feb. 2017.

VICINI, L.; SOUZA, A. M. **Multivariate analysis from theory to practice**, 2005. Santa Maria, RS: Federal University of Santa Maria. Available at: <http://w3.ufsm.br/adriano/livro/Caderno dedatico multivariada - LIVRO FINAL 1.pdf>. Accessed on: January 30, 2017.

VILLELA, S. M.; MATTOS, A. **A Hidrologia Apliada**. Mc Graw-Hi ed. Sao Paulo: Villela, Swami Marcondes; Mattos, Arthur, 1975.

YUNUS, A. P.; OGUCHI, T.; HAYAKAWA, Y. S. Morphometric Analysis of Drainage Basins in the Western Arabian Peninsula Using Multivariate Statistics. **International Journal of Geosciences**, v. 5, n. 5, p. 527-539, 2014. Available at: <http://file.scirp.org/Html/5-2800721_45490.htm%5Cnhttp://www.scirp.org/journal/PaperDownload.aspx?DOI=10.4236/ijg.2014.55049>. Accessed on: 09 Feb. 2017.

MIX
Papier aus verantwortungsvollen Quellen
Paper from responsible sources
FSC® C105338

Printed by Books on Demand GmbH, Norderstedt / Germany